RISKS
AND
CHOICES

RISKS
AND
CHOICES

Reza Noubary

To order additional copies of this book, contact:
Xlibris
844-714-8691
www.Xlibris.com
Orders@Xlibris.com
825251

Contents

ABOUT THE BOOK

This nation was built by men who took risks—pioneers who were not afraid of the wilderness, businessmen who were not afraid of failure, scientists who were not afraid of the truth, thinkers who were not afraid of progress, and dreamers who were not afraid of action.
—Brooks Atkinson

WE ALL TAKE risks. Individually, we may decide to go to college, get married, or change our job. These decisions are often made subjectively without carrying any quantitative analysis. Communities take risks when they decide to build a flood wall. Nations take risks when they join a treaty, invest in other countries, or go to war. The human race takes risk with its acceptance of the balance of nuclear power or climate change/global warming as a policy for peace. The acceptance of risk by individuals is often very different from the acceptance of the same type of risk by groups.

In sum, we are all exposed to some type of risk every day depending on the choices we make. Those who choose to sit on the sidelines and watch life pass by often take the bigger risk.

This book is about daily risk and is divide into ten chapters. The first eight chapters discuss certain aspects of risk and presents examples to enlighten the reader. Its focus is on general risk, risk-taking, and risk avoidance. It includes topics such as medical risk, investment, and environmental risk. The last two chapters (9 and 10) are devoted to methodologies that are used in earlier chapters and can safely be dropped by readers who are not interested in the mathematics behind the risk theory.

The author has experience in technical topics such as earthquake hazard assessment and risk analysis and has written two books and several articles where risks were evaluated formally.

Here are couple of points to remember while reading the book.

- Many great ideas have been lost because the people who had them could not take risk and stand being laughed at.–Unknown
- A ship in harbor is safe, but that is not what ships are for.–John A. Shedd
- Turtle makes progress only when it sticks its neck out.–James B. Conant

Also a poem regarding a particular choice.

Life is full of opportunities for temporary fun and joy
Once experienced most people are
encouraged to look for more
But if one think and assess its long-
term risks/consequences
He/She will realize that the return
is negative, that is for sure.
–Reza Noubary

ACKNOWLEDGMENT

I START WITH MY profound thank you to all individuals in my life who have guided me through the process of dealing with risk for their sage advice, help, and creative actions or suggestions, especially those who, like me, had to deal with many inflicted risks in their lives.

My gratitude to my wife, Zohreh, who does everything to comfort me to do my work. Hats off to my friends Steve Cohen, Youmin Lu, and Dong Zhang for their encouragement and support. I cannot fail to honor the memory of my mother and brother.

I would also like to acknowledge the contribution of my colleague Dr. JoAnne Growney, who worked with me on a couple of articles.

I DEDICATE THIS BOOK TO THOSE
WHO TAKE RISK TO MAKE THE
WORLD A BETTER PLACE.

ABOUT THE AUTHOR

T HE AUTHOR WAS born to an Azari family in early 1946. He was the youngest of a clan of eleven children, who, because of lack of access to medical care, died in early ages. His father was a police officer who struggled with drug addiction and alcoholism. His mother had no formal education as she was forced to marry at a very young age.

Although not easy, he managed to go through formal education and receive his BSc and MSc in Mathematics from Tehran University, and MSc and PhD from Manchester University in England. He worked in several universities in different countries. He has also been a visitor in Harvard, Princeton, U-Penn, UCLA, University of Maryland, University of Kaiserslautern, and Catholic University of Leuven. His research interests include risk analysis of natural disasters and applications of mathematics and statistics in sports. He is a fellow of the Alexander von Humboldt and a fellow or member of numerous professional organizations. He has published several scientific books and more than one hundred research articles in ten different disciplines. His outside interests include music, soccer, racquetball, and tennis.

He has experienced risk as an insider, outsider, majority, minority, winner, loser, believer, denier, single, married, student, teacher, son, father, grandfather, uncle, friend, and enemy. He has dealt with a hard childhood, poverty, health issues both physical and mental, revolution, war, shortage, and stress of learning new languages and adjusting to different cultures. He has two sons and a granddaughter and lives with his wife in a small town in rural Pennsylvania.

QUOTES ABOUT RISK

"If you risk nothing, then you risk everything." – Geena Davis

"In a world that is changing really quickly, the only strategy that is guaranteed to fail is not taking risks."– Mark Zuckerberg

"No risk, no reward. No pain, no gain." – Every sports coach.

"If you do not play you cannot win."– Judith McNaught

"Life is inherently risky. There is only one big risk you should avoid at all costs, and that is the risk of doing nothing."– Denis Waitley

"A ship in harbor is safe, but that is not what ships are built for." – William G.T. Shedd

"Two roads diverged in a wood … I took the one less travelled by, and that has made all the difference." – Robert Frost

"Security is mostly a superstition. Life is either a daring adventure or nothing." – Helen Keller

"If things seem under control, you are just not going fast enough."
– Mario Andretti

"Do one thing every day that scares you." – Eleanor Roosevelt

"Pearls do not lie on the seashore. If you want one, you must dive for it." – Chinese proverb

"And the day came when the risk to remain tight in a bud was more painful than the risk it took to blossom." – Anais Nin

"Take calculated risks. That is quite different from being rash." – General George Patton

"I can accept failure. Everybody fails at something. But I cannot accept not trying. Fear is an illusion." – Michael Jordan

"Opportunity dances with those on the dance floor." – Anonymous

"Yes, risk-taking is inherently failure-prone. Otherwise, it would be called 'sure-thing-taking.'" – Jim McMahon

"Progress always involves risks. You cannot steal second base and keep your foot on first." – Frederick Wilcox

"What great thing would you attempt if you knew you could not fail?" – Robert Schuller

"Taking a new step, uttering a new word, is what people fear most." – Fyodor Dostoevsky

"To dare is to lose one's footing momentarily. To not dare is to lose oneself." – Soren Kierkegaard

"You'll always miss 100% of the shots you do not take." – Wayne Gretzky

"Danger can never be overcome without taking risks." – Latin Proverb

"I will play it first, and tell you what it is later." – Miles Davis

"Take risks: if you win, you will be happy; if you lose, you will be wise." Swami Vivekananda

"To know what life is worth you have to risk it once in a while." Jean-Paul Sartre

"If you are not scared a lot you are not doing very much." Robin Sharma

"Sometimes, the biggest risks are those we take with our hearts." Anonymous

"If you want it, go for it. Take a risk. Do not always play it safe or you will die wondering." Anonymous

"Take every risk, drop every fear." Anonymous

"To win big, you sometimes have to take big risks." Anonymous

"If you are not willing to risk the unusual, you will have to settle for the ordinary." Jim Rohn

"If it is still in your mind, it is worth taking the risk." Anonymous

"When you take risks you learn that there will be times when you succeed and there will be times when you fail, and both are equally important." Anonymous

"Risk more than others think is safe. Dream more than others think is practical." Anonymous

"Trust because you are willing to accept the risk, not because it is safe or certain." Anonymous

"I am not where I am because of luck. I am where I am because I took risks others were not willing to take. The world rewards the risk-takers. It always has. It always will." Dan Pearce

"Ignore the risk, and take the fall. If it is meant to be, it's worth it all." Anonymous

"Be brave. Take risks. Nothing can substitute experience." Paulo Coelho

"If you do not take risks, you will never know." Anonymous

"Creative risk-taking is essential to success in any goal where the stakes are high." Gary Ryan Blair

"Take the risk or lose the chance." Anonymous

"When we stop taking risks, we stop living life." Robin Sharma

"If there is even a slight chance at getting something that will make you happy, risk it." A. R. Lucas

"Great love and great achievements involve great risks." Dalai Lama

"Find your dream, then risk everything to make it into a reality." Anonymous

"Creativity is the ability to take risks." Jim Kast-Keat

"To love is to risk. Therefore, to love is to be brave." Anonymous

"There can be no great accomplishment without risk." Neil Armstrong

"He who is not courageous enough to take risks will accomplish nothing in life." Muhammad Ali

"Life is all about taking risks. If you never take a risk, you will never achieve your dreams." Anonymous

"Sometimes life is about risking everything for a dream no one can see but you." Anonymous

"You have to take risks. We will only understand the miracle of life fully when we allow the unexpected to happen." Paulo Coelho

"It's better to cross the line and suffer the consequences than to just stare at the line for the rest of your life." Anonymous

"Turn a perceived risk into an asset." – Aaron Patzer, Mint Founder

"Risk comes from not knowing what you are doing." Warren Buffett

"If you are not willing to risk the unusual, you will have to settle for the ordinary." – Jim Rohn

"Taking risks does not mean you do not feel fear, acknowledge fear, or let fear inform you; you just do not let it stop you" – Caren Merrick

"I have not failed. I have just found 10,000 ways that would not work." – Thomas A. Edison

"There can be no vulnerability without risk. There can be no community without vulnerability. There can be no peace, and ultimately no life, without community." – M. Scott Peck

CHAPTER 1

What is Risk?

To know what life is worth you have
to risk it once in a while.
–Jean-Paul Sartre

USING A SIMPLE language, risk may refer to a situation involving exposure to danger or physical or emotional harm. It refers to an action or a decision whose outcomes are not fully predictable and its consequences are not completely controllable; things with possible negative effects that cannot be avoided but may be managed. Taking risk reflects our responses to loss and reveals a great deal about our personality and our intuition. We generally realize that we take risk not because we want to but because we have to. As such, we all need to know something about how to evaluate the risks involved in our decisions if we wish to make appropriate choices. Risk pervades virtually all areas of human endeavor, whether these endeavors are for personal, social, commercial, or national purposes.

The concept is complicated, and there is no universally accepted definition for it. Some commonly accepted definitions are the following:

- Risk is a situation or event where something of human value is at stake and where the outcome is uncertain.
- Risk is an uncertain consequence of an event or an activity with respect to something that people value.

Basically, these definitions express risk in terms of uncertainties and their consequences rather than quantities representing them. There is a general agreement that risk should be assessed using mathematical quantities so that we can use them for evaluation and comparison. Here are few considerations:

1. How do we define risk?
2. How do we measure (quantify) risk?
3. Whose views of risk should we consider?

    a. Individuals? Public? Experts?
    b. Are individuals rational when assessing risks?
    c. Are experts always right?
    d. Is perception of public reliable?

Other issues that require attention include risk estimation, risk management, and risk sharing (insurance).

Mathematical Expressions of Risk

Insurance: The business of risk

Let us start with the most general mathematical expression of risk–namely,

$$\text{Risk} = [\text{Probability \{event occurs\}}] \times [\text{Probable cost if event occurs}].$$

With this in mind, three types of risk models can be distinguished:

I. risk associated with the uncertainty of the occurrence of an undesired event and its fixed or deterministic consequences;

II. risk associated with the uncertainty of the magnitude of the consequences of the fixed or deterministic occurrence of an undesired event;

$$Risk = [Prob. \{event\} = 1] \times [P \{cost/event\}]$$

III. risk associated with the uncertainty of the occurrence of an undesired event and the uncertainty of the magnitude of its consequences.

Summarizing these, we may formulate them as

$$Risk = [Prob. \{event\}] \times [P \{cost/event\}];$$
$$Risk = Probability \times Severity.$$

From these definitions, it is evident that there are two basic components of risk:

1. a future outcome that can take a number of forms, some of them unfavorable;

2. a nonzero probability indicating that such unfavorable outcomes may occur.

Here are some other definitions involving these two components:

(A) Risk is the possibility of loss.
(B) Risk is uncertainty.
(C) Risk is the dispersion of actual from expected results.

(D) Risk is the probability of any outcome different from the one expected.

(E) Risk is the probability of an event times the cost (or loss) if the event occurs.

Each of us may prefer one over the others depending on our understanding, intuition, experience, and the type of the media we follow.

Risk and Rationality

Are human beings rational?

As pointed out, making individual decisions in the face of uncertainty reflects our responses to loss and reveals a great deal about our personality and our intuition about risk-taking. To see this, consider a case in which a company, as a bonus, is offering the employees to choose between two options:

Option A: 100 percent chance of winning $250.
Option B: 25 percent chance of winning $1,000, 75 percent chance of winning nothing.

The expected win (gain) is 250 x 1 = $250 for option A and 1,000 x 0.25 + 0 x 0.75 = $250 for option B. So the expected gain is the same for both options. However, 84 percent of employees chose option A, preferring to take a sure gain. A bird in the hand . . .

A year later, the company lost business to their competition and asked employees to share the loss.

Option A: 100 percent chance of losing $250.
Option B: 25 percent chance of losing $1,000, 75 percent chance of losing nothing.

The expected loss is $250 for both options. However, 80 percent of employees chose option B.

A related study puts these results in a different light through increasing the possible gains/losses and decreasing their probabilities.

Option A: 100 percent chance of winning $5.
Option B: 1/1,000 chance of winning $5,000.

The expected gain is $5 for both options. However, 75 percent of respondents chose option B, and their percentage increased to 85 percent when $5 and $5,000 where replaced by $1 and $1,000, respectively. This explains why people buy lottery tickets.

Option A: 100 percent chance of losing $5.
Option B: 1/1,000 chance of losing $5,000.

Expected loss is $5 for both options. But 80 percent of respondents chose option A.

So with small probability (1/1,000), we see different patterns. In example 3, the respondents believed the 1/1,000 chance of a gain was large enough and worth taking. In example 4, however, they believed the 1/1,000 chance of a loss was large enough to be avoided. Combining this finding with the previous examples, we have a general rule:

At least in a gambling context, people tend to be *risk seeking* to avoid a loss and *risk averse* to protect a gain. These examples demonstrate that our intuition about risk could be remarkably unreliable.

Lottery Example

In a similar study, the participants were offered an option of paying $1 for a 1/1,000,000 chance of winning $1 million. Nearly half of the respondents accepted the offer, including a large number of risk-averse individuals. The study also showed that a more extreme case (1/10,000,000 chance of winning $10 million) led to a significant increase in number of participants in general and risk averse participants in particular. This explains why lottery attracts so many people whose real reward is, in fact, a short period of a dream about a life with that much money. The bigger the prize, the bigger the dream, never mind how small the likelihood is.

Students at the University of Oregon, recruited by an ad in the student newspaper, were randomly given form I or form II of a questionnaire about vaccination. Form I described a disease expected to affect 20 percent of the population and asked people whether they would volunteer to receive a vaccine that protects half of the people receiving it. Form II described two mutually exclusive and equally likely strains of the disease, each likely to affect 10 percent of the population. The vaccine would give complete protection against one strain of the disease and no protection against the other.

Even though both vaccination forms proposed no risk to the volunteers and offered a reduction of the probability of the disease from 20 percent to 10 percent, the wording or "framing" of the situation lead to a "yes" response of 57 percent for form II but only 40 percent for form I.

A different study published in the *Scientific American* explores, at length, the role that the framing of a choice plays in the resulting decision. Compare, for example, the following pairs of choices:

Choices set A: Collect $50 for sure, or gamble with a 25 percent chance of winning $200 and a 75 percent chance of winning nothing.

Choices set B: Losing $150 for sure, or gamble with a 75 percent chance of losing $200 and a 25 percent chance of losing nothing.

For A, most people preferred not to gamble; whereas for B, most preferred to gamble even though the expected outcomes are identical.

Finally, here is an interesting example of how we think about risk-related situations. The winning ticket in a lottery was 865304. Three individuals compare the ticket they hold to the winning number. John holds 361204; Mary holds 965304; Bill holds 865305. All three lost since the chance of losing was only 999,999 in a million. Why, then, was Bill very upset, compared to Mary and John, who were only mildly disappointed?

References

Slovic, Paul, Baruch Fischhoff, and Sarah Lichtenstein. "Behavioral Decision Theory Perspectives on Protective Behavior." https://scholarsbank.uoregon.edu/xmlui/bitstream/handle/1794/22331/slovic_250.pdf?isAllowed=y&sequence=1.

Media Effect

Studies have found that people's attitudes toward risk are often formed based on the coverage of these risky events in popular media and their local newspapers. For example, some people think that traveling by plane is inherently more dangerous than driving because plane crashes are often headline news. To support the idea, one study examined the coverage by newspapers in New Bedford, Massachusetts, and Eugene, Oregon, and found the following patterns: Although diseases take about sixteen times as many lives as accidents, the newspapers contained more than three times as many articles about accidents. Although diseases claim almost one hundred times as many lives as do homicides, there were about three times as many articles about homicides than disease-related deaths. Furthermore, homicide articles tended to be more than twice as long as articles reporting disease and accident deaths.

The people who read these newspapers assessed the risk of death by homicide as much greater than the risk of death by accident, which, in turn, was much greater than the risk of death by disease. Although this assessment was incorrect, it was a correct interpretation of the deaths about which they read. In other words, the readers' number sense was working correctly on bad information.

Another study investigated the ordering of the perceived risk by two groups of ordinary citizens and experts regarding certain risky activities and technologies. Out of twenty such activities, ordinary citizens ranked nuclear power as the riskiest, whereas experts ranked it the least risky. The deviation from the expert's ordering is believed to be partly due to the news coverage and the group membership.

A Working Definition of "Risk"

We purchase fire insurance to protect ourselves against the consequences of a severe loss. In deciding to insure against the risk of fire, many of us consider only the magnitude of the possible loss and pay little attention to its probability. In fact, we may have no estimate of the probability that a fire loss will occur. In some other situations, we may equate "risk" with "probability," as in the following statements:

> The risk of a broken leg is greater for an inexperienced skier than for an experienced skier.

> The risk of death or injury in automobile accidents is greater for those not wearing seat belts than for those who are.

Lack of a fixed definition for risk may not be troublesome in our individual decision-making, but it can be a hidden source of disagreement when groups are comparing the magnitudes of various risks.

Those scholars and scientists who practice systematic risk assessment stipulate the following: Risk deals with the uncertainty of a possible loss or damage, but the risk is not the same as the uncertainty, nor is it the same as the loss. A correct characterization of risk involves both. The *risk* of a particular undesired event is the *expected loss* associated with the occurrence of the undesired event. In mathematical terminology, this characterizes risk as an expectation (i.e., a product of a probability and a numerical consequence, if numerical values are available.)[1] A formula for calculating the risk associated with a particular undesirable event is

Risk = Probability of event x Probable cost if event occurs.

For example, if the probability that our $80,000 home is destroyed by fire in a given year is 1/2,000, or 0.05 percent, then

Risk due to fire (per year) = (0.0005) x ($80,000) =$40.

This calculation shows how the probability and the cost/outcome are combined to give a measure of the "risk."

Investment Risk

Financial advisors often try to evaluate investor's tolerance for risk/loss. But, again, the question is, what do they mean by risk? Numerous factors affect markets and investing and requires basic understanding of the risks involved. The risk of a security has two main components: unique risk and market risk. Investors with a low tolerance for risk try to reduce or avoid it by actions such as *diversification*. Unique risk can be *reduced* or even *eliminated* by holding a well-diversified portfolio. Market risk associated with market's wide variations cannot be removed by individual action. As such, the risk of a fully diversified portfolio is the market risk or its sensitivity to market changes. This sensitivity is generally measured by a quantity known as beta (β).

A security with $\beta = 1.0$ has average market risk; a well-diversified portfolio of such securities has the same risk as the market index. A security with $\beta = 0.5$ has below-average market risk—a well-diversified portfolio of these securities tend to move half as far as the market moves and has half the market's risk. For the security with $\beta = 1.0$, if the market index changes 20 percent, its return (or price) would change 20 percent too. For the security with $\beta = 0.5$, if the market index changes 20 percent, its return (or price) would change 10 percent. For the security with $\beta = 2.0$, if the market index changes 20 percent, its return (or price) would change 40 percent. The regression analysis provides an effective approach for finding β.

A Little History

Risk can be associated with many facets of life from health risks to business risks to risks related to natural disasters. As risk theory is examined, it becomes evident that, often, people have different perceptions of risk; what may seem catastrophic to some may be of only minor concern to others. With these discrepancies, it is necessary to see how risk is actually defined and then question if a uniform definition even exists. This discussion will then lend itself to four different methods for handling risks. Throughout this report, examples relating to the risks associated with natural disasters and business ventures will be used. Finally, some of the mathematics behind risk assessment will be reviewed.

In discussing risk, there is often difficulty because, as mentioned earlier, individuals have different perceptions of risk. For example, one individual might consider hurricanes to be the riskiest natural disaster, while others might consider earthquakes to be the riskiest. Surely, a Californian would be likely to follow the latter line of thinking or even forest fires in place of earthquakes. A variety of classifications for risk could be given to help reveal how one labels something to be "risky." The combination of some of the following factors may play a role in risk classification. Some of the factors are whether something is catastrophic or minor, controllable or uncontrollable, and direct or indirect. Other characteristics that are often evaluated are whether the event is familiar or new, equitable or inequitable, internal or external, and temporary or permanent. Still other items that are considered in classification are whether something is expected or unexpected, fatal or nonfatal, general or specific, interactive or independent, instantaneous or gradual, and, finally, whether the individual is affected voluntarily or involuntarily. Some suggest that knowledge theory can explain some perceptions.

A survey was done to see why people have different perceptions of risk. The population sample included college students, a league of women voters, and experts. These people were asked to rank a list of technologies and activities according to the risk associated with each one. The survey included items such as nuclear power, motor vehicles, swimming, firefighting, large construction, motorcycles, and commercial aviation. Interestingly, the majority of college students and league of women voters ranked nuclear power as most risky. In contrast, motor vehicles were at the top of the experts' list as most risky. Eighty-four percent of college students and league of women voters answered similarly; while only 35 percent of the league of women voters and the experts has matching answers.

The media also plays its role in shaping perceptions of risk. Society often makes the association that what happens in the news is the reality of a situation. So when airplane accidents are given more news coverage than automobile accidents, conclusions are often drawn. Individuals may tend to conclude that airplane accidents occur more frequently than car accidents, or better yet, that airplane travel (commercial) is riskier than travel in motor vehicles. In reality, more automobile accidents occur annually than airplane crashes. The above-mentioned survey also provided the expert ranking that motor vehicles are indeed riskier. In another case, it is found that newspapers typically print more articles about death by homicide than death by disease. As a result, readers might assess the risk of death by homicide as greater than the risk from disease, when in fact the opposite is true.

When trying to uncover what people mean when they say that something is risky, we begin to wonder, "What exactly is risk?" The following are some of the common textbook definitions identified: Risk is the chance of loss. Risk is the possibility of loss. Risk is uncertainty. Risk is the dispersion of actual from expected results. Risk is the probability of any outcome that is different from the expected one. Risk is the condition in which a possibility of loss exists. Risk is the potential for the realization of unwanted negative consequences of an event. Risk is the probability of an undesired event times a projected cost of the negative consequence if the undesired event would occur. Risk for business ventures is the beta value for a particular security.

These definitions have varying degrees of usefulness. In fact, most individuals have difficulty providing a single definition for risk. Some of the underlying ideas behind most of these definitions are the existence of some uncertainty about an event and the potential for loss, whether it is financial, material, or otherwise. Without the existence of a commonly accepted definition of risk, often, researchers of risk have different frames of reference. This is not to say that we necessarily need a singular definition, but it must be acknowledged that there is some disagreement. Some of the definitions given seem rather vague using words such as "uncertainty" and "chance," while other definitions lend themselves rather well to mathematical interpretation. For example, risk as the dispersion of actual from expectation is better known to mathematicians as variance. One might be familiar with expected return, which is the mathematical name for the definition that begins with risk is the probability of an undesired event times its consequences. The mathematical assessment of risk will be addressed in more detail later.

Dealing with Risk

Avoiding one type of the risk often leads to inviting another type.

People at risk often handle it differently. The four major methods for handling risk are avoidance, retention, reduction, and transfer (insurance). Avoidance is when an individual realizes that a risk is present and tries to avoid it. The individual does not want to suffer the negative consequences of an event, and, therefore, he or she does not participate. The person tries to remove himself of herself from the risky situation. For example, to avoid losing money in a business venture, simply do not get involved at all. To avoid losing money on the stock market, simply do not buy stock. With the case of natural disasters, avoidance is slightly more difficult. The avoidance of one type of disaster may be possible by moving to a different location, but in doing so, he or she may be exposing himself or herself to other disasters. For example, people can avoid the risk of a flood by relocating to a non/no flood plain region. Unfortunately, the new region may be prone to hurricanes or earthquakes. With Mother Nature ultimately in control, avoidance is not always the best solution, particularly when done by means of evacuation. This procedure is costly and relies heavily on prediction techniques that are done with a degree of uncertainty.

Another option for handling a risk is retention. People can voluntarily retain the risk in some instances, but in other cases, it is involuntary. Regardless of whether it is voluntary or involuntary, risk retention means that the individual is responsible for loss if it occurs. When people agree to retain the risk, they are accepting to bear the consequences if the undesired outcome results. For example, in business ventures where stock is purchased, the risk is losing money. If a person willingly purchases stock, then he or she has practiced risk retention. That means that the person must bear the losses that may be suffered by undesired drops in stock prices. With regard to natural disasters, risk retention is practiced by most people. Without a form of insurance to protect people from natural disasters, most people simply hope that they will not suffer loss because of a natural disaster; but if a loss does occur, they are responsible. In some cases, government relief money has been allocated to help bear the loss of some disasters such as flooding, but too often that is the exception, not the norm.

A third method for handling particular risks is risk reduction, which is often associated with preventative measures that will reduce the chances of the negative consequence from occurring and reduce the losses if the event does occur. Preventative measures can be easily described using the example of health risks. People have begun to eat foods lower in fat and cholesterol to reduce their chances of suffering a heart attack. In the business world, investors can help to reduce the risk of their stocks by purchasing mutual fund, a portfolio of stocks that is a collection of stocks that have varying measures of riskiness. Buying several stocks rather than a single stock provides a balance in case one stock may drop dramatically, we can only hope that one of the other stocks would increase considerably.

Risk reduction for natural disasters has been the topic of great interest since the beginning of this decade because the United Nations declared the 1990s as the International Decade for Natural Disaster Reduction. Considering that a natural hazard in and of itself is not problematic, we realize that problems do arise when humans enter the picture. If the natural phenomenon such as earthquakes, volcanoes, landslides, and hurricanes occurred in uninhabited areas, they would be of little societal consequences. With a quadrupling of the world's population during the twentieth century, hazards are striking heavily populated areas and disasters are becoming more frequent and more severe. Recall, the 1954 flood in China resulted in approximately forty thousand deaths, and similarly, in 1988, an earthquake rumbled in what used to be USSR and the death toll was well above forty thousand.

Risk reduction for natural disasters can be achieved through engineering technologies and improvements in the structures of buildings and infrastructures. In California, there are certain building codes that must be met before a building will pass inspection, and some of these pertain to making buildings "earthquake proof." This measure actually relates to reducing the consequences of a disaster. The Decade has certain goals that are described as, "It aims to reduce through appropriate action, the loss of life and property damage due to natural disasters." This will be a global effort and encourages people to work together. Such a project will test our ability to accept that people have different perceptions of risk, particularly when funds are allocated or research is done for specific hazards that some (but not all) people face. For example, consider the problems that may arise when the people of Bangladesh want the problem of flooding addressed while Californians want earthquake issues addressed. To achieve the goals of the Decade, we must all be sensitive to one another's motives and beliefs, which may have a strong impact on our attitudes toward risk and risk reduction.

Looking back, we can try to reduce the risk or reduce the loss, or both. More often than not, with regard to natural disasters, reducing the loss is the best we can hope to do, since, once again, Mother Nature is ultimately in control.

The final method for handling risk is risk sharing, which is often referred to as insurance. The basic concept behind insurance is that many people face a situation where there is the potential for an undesired outcome. Ironically, only a select few will actually suffer the undesired consequence or loss. To help share the loss and make it easier for the few to bear, the entire group concerned pools their money together (collected by an insurance company), and then it is redistributed. The money is handed out as needed to those who actually suffered a loss. Today, there are many different types of insurance available. Everything from health insurance to life insurance to fire insurance to car insurance to earthquake insurance can be purchased in some areas. For example, in

California, on September 21, 1990, Governor Deukmejian signed into law a bill that mandates earthquake insurance for all residential property owners. In flood plain regions, such as Bloomsburg, Pennsylvania, flood insurance can be bought from an insurance company through the National Flood Insurance program in specific parts of the United States. Unfortunately, natural disasters are not covered under traditional homeowners' insurance. Although it seems that for additional charges, some other disasters can be insured.

To help share the losses resulting from natural disasters, proposals have been made suggesting a type of international insurance for disasters. This is a hopeful idea that I do not expect to see materialize for many, many years. Part of the problem is a lack of information about past disasters. The historical records are incomplete to make accurate estimations about the predicted amount of disasters in a given time frame. With the help of mathematics, some of this information is becoming available as models are being formulated to fit the data of the past. Another problem is that global warming is rendering some past statistics somewhat absolute.

Considering risky business ventures, risk sharing can be accomplished by having several investors involved in an investment. As a result, any loss will be divided among all the investors. Unfortunately, just as all of the losses would be shared, so, too, would all the profits be shared.

The handling of risk often goes hand in hand with risk assessment. Intuitively, we can agree to retain the ones that are not too risky. The description that one option is riskier than another has a better chance of acceptance if the risks involved could be quantifiable. In fact, there are several ways for quantifying risk. The simplest measure for risk is expected value (also called expected return). This defines risk as

Risk = (Probability of undesired event) x (Cost of loss).

If the expected value for two situations is calculated, then a conclusion can be made about the comparison of risk between the two. The condition that yields the larger expected return is less risky than one with a smaller expected return. Consider the following two lottery options:

Lottery A: 50 percent gain $100; 50 percent lose $30
Lottery B: 60 percent gain $150; 40 percent lose $80

To evaluate which lottery option is riskier, calculate the expected return.

$$E(A) = (0.5)(100) + (0.5)(-30) = \$35$$
$$E(B) = (0.6)(150) + (0.4)(-80) = \$58$$

After comparing the expected returns, we can assess that the first lottery is riskier than the second one; therefore, the second is less risky than the first one. Looking at another example, we can see the weakness of using the expected value. Consider there is a 50 percent chance of losing $1,000 and a 50 percent chance of gaining $1,000. In another situation, there is a 50 percent lose $1 and 50 percent gain $1. Looking at the expected returns, they are the same:

#1: (0.5)(1,000)+(0.5)(-1,000)=0
#2: (0.5)(1)+(0.5)(-1)=0

In a real-life situation, most people would agree that there is more risk in possibly losing $1,000 as opposed to $1.

Realizing the shortcomings of this method, there is another way to calculate risk using what is known as variance. Variance is defined as the dispersion of actual from expected results. This can be more easily explained as how much the data varies from a given point. Looking at business dealings involving stocks, the variance of a stock describes how much the stock fluctuates. If a stock has a large variance compared to another stock, then the initial stock is said to be riskier than the second. Having a large variance implies that the stock has a past record of increasing in price and then decreasing considerably, and this pattern of change continues to persist again and again.

Calculations can also be made to assess risks apart from the business world. Particularly the risks associated with earthquakes can be examined. When trying to evaluate earthquake occurrences, it is useful to predict the number of earthquakes that will occur in a particular area during a specified period based on the historical records of the fault. This can be done using the Poisson process because the occurrence of earthquakes follows the Poisson distribution.

Overall, risk assessment can be paramount in evaluating different situations and decision-making. It does not dictate or prescribe a particular behavior but helps to more clearly illustrate a circumstance to those studying the risk of a particular situation.

Interestingly, people continue to use personal preferences to make decisions even when situations are deemed to be very risky. This may seem ironic, but some people are attracted to risks more than other people. This is not to say that one is better or worse than the other. Risk assessment simply allows for better-educated and informed decisions.

In conclusion, risks are everywhere and cannot be entirely ignored. We must try to expand our ideas in an attempt to understand other people's perspectives and work toward addressing those risks that affect not only our lives but also the lives of other people. We must make a conscious effort to handle risks appropriately and try to reduce when the situation calls for such action. The actions of those in science and technology have led to advances in prediction and risk assessment, which have helped to save lives. I hope that people will become more aware of the risks that they take or the risks that they face.

Public Views of Risk

In this section, different expressions of risk together with policies for adapting risk, quantifying risk, and public views of risk are briefly discussed. Topics include policies for adapting to risks, perceptions of risk, quantifying risk, mathematical expressions of risk, risk assessment, assessment by words, public views of risk, medical risks, and life expectancy criterion.

When specifying acceptable levels of risk, the public and their political representatives often have a different perception of the relative risks from what is supported by the facts. For example, nuclear power is considered a major risk by members of the public; whereas, according to experts, the risks are much less than those associated with swimming, cycling, or travelling by train.

One implicit distinction made in the public assessment of risk is between *voluntary risk* and *involuntary risk*. We all accept voluntary risk when crossing the road or driving a car. The advantages are thought to justify the risk. However, the risks and the consequences suffered by the people of Bhopal in India when 2,500 people died as a result of leaks from the Union Carbide factory are *universally condemned*.

Some Risks Are Inevitable

Remember the saying that not to take a risk is taking a risk of a different kind. In the field of medicine, whooping cough inoculations have saved many lives, but there is a finite risk in having the vaccination. With many new drugs that save lives, there is also some risk of damaging side effects. Many people choose to undergo major surgery; aware that there is a one-in-a-hundred chance of something going seriously wrong, because they judge or are advised that without it, their risks are much greater. The insect killer DDT is now generally deplored, but when banned in Sri Lanka in the early 1960s, there was a raging and virulent epidemic of malaria and many people died unnecessarily because the means of controlling mosquitoes was prohibited.

Role of Statistician

Statisticians should try to make available to decision makers the best possible estimates of risks. They should make them aware of the limitations of the estimates and of the need for better monitoring procedures. They should be able to expose and criticize bad emotionally based quasi-scientific estimates of risk. But they should not expect decisions to be made on the basis of their estimates alone; there are other legitimate factors.

Policies for Adapting to Risks

*Man cannot discover new oceans unless he
has the courage to lose sight of the shore.*
–Andre Gide

According to experts, there are three very general categories of response to risk that may be identified as follows:

(a) Survival of the species, without regard for the individual. This is a natural process–plants and animals disperse and propagate freely; only those members of the species that are best adapted to the extremes of the new environment will survive. The individuals of the species are unaware of the risks that they take.

(b) The recognition that disasters occur, but a belief and acceptance that they are unavoidable. Often, such disasters are attributed to the wrath of the gods who may be appeased by suitable sacrifices. This is called fatalism. Poor people throughout history have adopted this philosophy of necessity.

(c) Recognition that there is a quantifiable risk of disaster, accompanied with the assertion that these risks can be reduced, and perhaps eliminated. This is the *modern* approach: today, people turn for advice to statisticians rather than to priests! Of course, one may choose not to protect against a disaster if the costs of protection are too great, and one may deliberately choose to take risks if the potential gains are great enough. These choices may be very difficult to make, but there are recognized rules of cost-benefit that may assist. It is required that

Cost of avoiding risk < Risk times consequence.

Although the responses of people making decisions will vary, there are some common factors that we may consider. However, it would be wrong to suppose that all decisions are made on a rational quantitative basis; there are often other pressures on decision makers.

Example: Individuals exhibit a range of reactions to natural hazards and display a range of knowledge. Many people are unaware that they are at risk unless there has been a recent flood. When they are aware, they seem to start with some anchor belief about the probability of a flood and then modify this belief with experience.

Most people have a hard time interpreting small probabilities and refuse to consider those below a threshold. For floods with a return period greater than ten years, migration means that many people will not have experienced such a flood and might be ignorant of the possibility. Those who have themselves or their friends experienced such a flood were more willing to buy insurance or protect their homes than people who have not had experience. Taking advantage of the range of information are unscrupulous builders or current building owners who can find ignorant customers. Thus, it is often argued that for floods with return periods of a century or more, the level of popular ignorance is likely to be complete, or most people will not choose to think about such floods.

Risk Assessment

Few people would question the value of risk assessment.

Risk deals with the uncertainty of a possible loss or damage. To take an action involving a risk is to take a chance or a gamble; it implies a degree of uncertainty and inability to control fully the outcomes or consequences of such an action. Do we have to take risk? The answer is yes. In fact, not to take a risk may mean taking a risk of a different kind. It is the general agreement that progress is impossible without taking risks. Unfortunately, most definitions of risk found in the literature use incorrect or vague terminology, which leads to misunderstandings. Often, the terms "risk" and "probability" are used indiscriminately.

When calculating and specifying probabilities for risk assessment, there are some basic requirements:

(a) They must be expressed in clear language that is understood. "Safe as the Bank of England" will not do. We need either to give the value of a parameter that has a specified probability such as 1:1,000, or, alternatively, the probability of a particular specified event. In terms of, for example, flood protection, this could be the probability of exceeding the level of existing protection walls.

(b) The probabilities must be stated in terms of the chance of some hazard occurring in a year, for each one hundred miles driven, or each time some activity (such as a free fall parachute jump) is undertaken.

(c) Some recognition of the confidence that can be placed in these estimates should be included. Statistical estimates of errors may be quoted, but these are not always possible. At the very least, the extent and type of the data on which the estimates are based should be specified.

Assessment by Words

When separating out uncertainty, the probability is sometimes expressed in verbal terms rather than as a number. This approach can be challenged because words are useful to convey meaning, only provided that the writer and the reader (or speaker and listener) agree on the meanings to be ascribed to the words. However, in the realms of describing uncertainty, words do not have a generally accepted and agreed meaning.

As an illustration, the following list of ten expressions was all culled from the same substantive article that was discussing, in a literary rather than numerate style, some forecasts that had been made in the consumer durables field:

Probable	Hoped	Expected
Quite certain	Possible	Doubtful
Unlikely	Not unreasonable that	Not certain
		Likely

Some 250 executives on middle and senior general management programs at the London Business School and elsewhere were asked to rank these ten words or phrases in decreasing order of uncertainty. The table below summarizes the results obtained, the expressions being reordered in the table in descending order of average rank. The final column, giving the range of ranks given to each of the ten expressions, illustrates the considerable overlapping of ranks for many of the expressions among the respondents and thus the inconsistent use being made of these words. Indeed, *only three out of the 250 respondents produced precisely the same rankings*. A further experiment, with a smaller group of respondents, repeated the ranking process after an interval of about three months and demonstrated that individual respondents *are not even consistent over time* in their ranking of the same expressions.

Table: Ranking of Uncertainty Expression

Expressions	Average Rank	Range of Ranks
Quite certain	1.10	1-3
Expected	2.95	1-6
Likely	3.85	2-7
Probable	4.25	2-9
Not unreasonable that	4.65	3-7
Possible	6.10	3-9
Hoped	7.15	3-10
Not certain	7.80	3-10
Doubtful	8.60	7-10
Unlikely	8.75	3-10

CHAPTER 2

Some Examples

*I am having a great time. I wish you
were her. Sorry I meant here*

This chapter includes some everyday examples involving risk.

Pandemics

P ANDEMICS ARE LARGE-
SCALE outbreaks of infectious
disease that can greatly increase morbidity and mortality over
a wide geographic area and cause significant economic, social,
and political disruption. Evidence suggests that the likelihood
of pandemics has increased over the past century because of
increased global travel and integration, urbanization, changes in
land use, and greater exploitation of the natural environment.
These trends likely will continue and will intensify.

Risk Factors

Pandemic risk is driven by the combined effects of spark risk (where a pandemic is likely to arise) and spread risk (how likely it is to diffuse broadly through human populations). The foci of both risk factors often overlap; it is measured by the estimated probabilities that, in any given year, pandemics of varying degrees of severity will occur. Expected annual losses defined in the probabilistic sense as the sum, across severities, of the losses associated with a pandemic of any given severity multiplied by the probability that a pandemic of that severity will occur in the coming year. Other factors involved are loss (the consequences of a pandemic, in terms of lost income or lost lives) and costs (the expenditures made to prepare for—or recover from—a pandemic.) Pandemic severity is the excess death attributable to a given pandemic.

Flying versus Driving

Some people think that traveling by plane is inherently more dangerous than driving a car. According to the National Safety Council, during the life of a person, the odds of dying in a motor vehicle accident versus air and space transport are 1 in 98 and 1 in 7,178, respectively. This indicates that flying is far safer than driving. However, for some, flying may feel more dangerous because our perception about risk is usually formed based on factors beyond mere facts. For example, most people think that they are good drivers and, as such, feel safer because driving affords personal control. Additionally, a car crash does not often lead to death; whereas, plane crashes are often catastrophic. It kills many at once and grabs the attention of major media, which make people more aware of plane accidents as compared to car accidents. In what follows, a detailed analysis is presented.

Measuring Risk

In general, there is a lot more to calculating and comparing risk than one might think. According to the experts, for an average American, the annual risk of being killed in a plane crash and a motor vehicle are about 1/11.000,000 and about 1/5,000, respectively. What does this mean to you and me? First, most of us are not the average American. Some people fly more than others, while some do not fly at all. So if we take the total number of people killed in commercial plane crashes and divide that into the total population, the result gives a good general guide, but it is not specific to our personal risk. Here are other useful numbers we may prefer to use:

1. Risk per person: dividing the number of people who die into the total number of people.
2. Risk per flight: dividing the number of victims into the total number of flights passengers took.
3. Risk per mile: dividing the number of victims into the total number of miles all of them flew.

Example

In 1995, out of every one hundred million people, about 16,300 died in automobile accidents and 111 were killed in commercial flight accidents. The number of deaths per one hundred miles were, respectively, 3 and 100 for one hundred million miles traveled and 30 and 20 for every one hundred million trips made.

This shows that the risk of death per mile is 33 times higher for car. However, the risk of death per trip is about 1.5 times higher for airplanes.

Discussion

The article "How Risky Is Flying" by David Ropeik provides an interesting discussion on air travel versus car travel. All the above calculations are useful and accurate. However, the one most relevant to us depends on our personal flying patterns. Some of us take many short flights. Some fly less but longer flights. Since, according to the available data, majority of plane crashes take place in connection with takeoffs and landings, one may conclude that the risk is more a matter of how often one flies and less a matter of how far. If you are a frequent flier, then the risk per flight means more to you. For people who fly occasional long distance, the risk per mile means more. A frequent long-distance flier may consider both.

Ropeik also states that like other countries, the number of plane crash fatalities in the United States varies widely from year to year. So the calculation of risk based on one year versus average of five, ten, or twenty years may vary significantly. In some years, no plane crashes occur, or at least very few do. This makes the value of the risk per year misleading. If we average things over five or ten years, some other factors will muddy the waters. In the last five years, safety factors have changed. A ten-year average might be misleading too.

According to Ropeik, despite all that has been pointed out, numbers are a great way to put risk in general perspective. Whatever the measure we use, flying is less risky than traveling by cars. However, numbers are not the only way, nor are they even the most important way we judge what to be afraid of. Risk perception is not just a matter of the facts. For example, consider the risk awareness factor. The people who are more aware of a risk become more concerned about it. This explains why when there is a plane crash in the news, flying seems scarier to many of us, even though that one crash has not changed the overall statistical risk significantly.

An article titled "Road Rage Statistics Filled with Surprising Facts" shared the following interesting statistic regarding road rage:

- A recent data from the National Highway Traffic Safety Administration (NHTSA) shows that 94 percent of all road accidents are caused by driver error. Out of those accidents, 33 percent could be linked to behaviors such as road rage, illegal maneuvering, or misjudging the intent of another driver.
- In 1990, the AAA Foundation for Traffic Safety studied more than 10,000 traffic accidents that were related to driver violence. They found that over a seven-year period, more than 12,500 injuries were linked to these acts. Road rage was also the cause of 218 deaths, mostly conducted by angry drivers. Surprisingly, this number has been steadily increasing at a rate of 7 percent each year. In fact, data gathered by the website *Safe Motorist* indicates that 66 percent of recent traffic fatalities were due to aggressive driving. Also, more disturbingly, 37 percent of those fatalities were caused by use of a firearm rather than a usual accident. This shines a light on the fact that road rage often does not end once a driver is off the road or outside their car. Half of drivers who shared road rage stories admitted to engaging in aggressive behavior in response.
- Finally, in recent years, the use of cellphones to talk or text has become a great contributing factor to the accidents of all kind.

References

http://drivingschool.net/road-rage-statistics-filled-surprising-facts/.

Ropeik, David. "How Risky Is Flying." http://www.pbs.org/wgbh/nova/space/how-risky-is-flying.html.

"Road Rage Statistics Filled with Surprising Facts." https://
   drivingschool.net/road-rage-statistics-filled-surprising-facts/.
"How Risky IS Flying? It Depends." https://bigthink.com/
   risk-reason-and-reality/how-risky-is-flying-it-depends.

Drug Overdose

Drug overdose has become a serious problem in the United
States and around the world. According to the Centers for
Disease Control and Prevention (CDC), on average, ninety-one
Americans die every day from overdose of opioids such as heroin,
the synthetic opioid, fentanyl, prescription painkillers such as
OxyContin, and now the elephant tranquilizer, carfentanil. Drug
use is now the leading cause of accidental death in America. In
2016 alone, more Americans (around sixty thousand) died from
drug overdoses than of AIDS, guns, or car crashes in a single year.
In 2015 statistics, Pennsylvania, with 3,264 deaths, was ranked
only second to California, with 4,659 deaths. This is significant,
noting that the population of Pennsylvania is less than one-third
of the population of California.

According to the estimates, in the United States alone, more
than twenty million people aged twelve and older struggle with
a substance abuse disorder, of which two million struggle with
addiction to prescription painkillers and six hundred thousand
suffer from heroin addiction. Since 1999, the amount of prescription
opioids sold on the market has quadrupled.

Overdose

An overdose is defined as the intentional or accidental ingestion
of a drug over the normal or recommended amount. The body's
response to it is often accompanied by severe symptoms as it is
overwhelmed and is unable to metabolize the drug quickly enough.
An overdose can cause a person to fall into unconsciousness, enter
a state of psychosis, or experience painful symptoms. Each type

of overdose poses significant health risks, including contributing to a person's death.

History

Most people relate opioid epidemic to illegal drugs only. In fact, it really started with legal drugs. Back in the 1990s, doctors were persuaded to treat pain as a serious medical issue. Knowing this, pharmaceutical companies took advantage of it. Through a big marketing campaign, they got physicians to prescribe products like OxyContin and Percocet in droves—even though the evidence for opioids treating long-term chronic pain was very weak and the evidence that opioids cause harm in the long term was very strong.

Thus, painkillers proliferated, landing in the hands of not just patients but also teens rummaging through their parents' medicine cabinets, and family members and friends of patients, resulting in their availability on the black market. Subsequently, opioid overdose deaths trended up—sometimes involving opioids alone, other times involving drugs like alcohol and benzodiazepines typically prescribed to relieve anxiety.

According to the article by German Lopez, "In one year, drug overdoses killed more Americans than the entire Vietnam War did." Having noticed the rise in opioid misuse and deaths, officials have cracked down on prescription painkillers. Law enforcement, for instance, threatened doctors with incarceration and the loss of their medical licenses if they prescribed the drugs unscrupulously. Though helpful, many people who lost access to painkillers were still addicted. So some who could no longer obtain prescribed painkillers turned to much cheaper, more potent opioids: heroin and fentanyl, a synthetic opioid that is often manufactured illegally for nonmedical uses. This pushed the number of victims to a new high. In addition, according to a 2016 report by the surgeon general, only 10 percent of Americans with a drug use disorder obtain specialty treatment, which is partly due to the shortage of treatment options.

Worldwide Statistics

According to the international bodies, in 2014, there were about 207,000 drug-related deaths, with overdose accounting for up to a half of them and with opioids involved in most cases. China accounted for 49,000, and Oceania (including Australia and New Zealand) had around 2,000. In European Union countries, more than 70,000 lives were lost to drug overdoses in the first decade of the twenty-first century.

Final Words

It seems that overdosing has taken the place of moderation in our culture. Every day, people are dying from overdose, and the numbers are growing at alarming rates. According to the reports, for opioid overdoses, the three-year period between 2014 and 2016 has been the deadliest. Some explanations are the following:

1. The cost of heroin is roughly five times less than prescription opioids on the streets.
2. Painkiller addiction (people who abuse or who are dependent on prescription painkillers are forty times more likely to abuse heroin).
3. Treatment options are few or nonexistent.

Considering these, we need to

- look at the problem deeper, especially the social and behavioral side of it;
- investigate why about one hundred million Americans suffer from chronic pain despite the fact that we spend more money for health care than the next ten big spenders combined; and
- seriously pursue policies that would curb this growing problem.

References

Lopez, German. "In One Year, Drug Overdoses Killed More Americans Than the Entire Vietnam War Did." https://www.vox.com/policy-and-politics/2017/6/6/15743986/opioid-epidemic-overdose-deaths-2016.
https://alcoholrehab.com/drug-addiction/overdose-facts-myths-and-symptoms/.

Fake News

Fake news is based on information that is not correct. It is often created by individuals who want to spread fear and panic, or by people who believe in theories that are not based on any scientific fact. In an age where the Internet is frequently the main source of information, news audiences are at higher risk than ever of encountering and sharing fake news. Every day, consumers all over the world read, watch, or listen to the news for updates on everything.

Fake news is nothing new—we have long been exposed to propaganda, tabloid news, and satirical reporting. But now, with the dependence on the Internet, promotion of trending stories on social media, and new methods of monetizing content, we have found different ways to relay information without using traditional media outlets. A single story posted on a personal or biased website can go viral and lead to additional content that is distorted. The original authors may be fully aware that the story is a lie, but with no throttling or inspection of content, the story can take on a life of its own, go viral, and spread misinformation while also leaving a tarnished impression of legitimate media. Regardless of how far the story spreads or your belief in its contents, fake news stories present significant risk to people, industries, and governments. For example, if no one believes what they read—or more precisely, only believe what they want to believe—democracy suffers. Younger consumers are often at greater risk of exposure to fake news than older generations by sheer virtue of their higher social media usage.

Lies

**Our lives improve only when we take chances,
and the first and most difficult risk we can
take is to be honest with ourselves.**
–Walter Anderson

We live during a time in which now, more than ever, spurious realities are manufactured by the media, politicians, and political groups. Major news agencies present selected news and analyses based on their version of "truth" and leave readers and viewers wondering what to believe or whom to trust. Of course, what is happening is not entirely new. Deceptions have been used by many groups, including politicians and the media, since the dawn of Western civilization. The history of humankind is full of well-known lies and crafty, seasoned liars. As Napoleon Bonaparte said, "History itself is nothing but a set of lies agreed upon." Even scientists who are supposed to pursue the truth have been shown to deceive.

Role of the Lies

According to experts, deceit and falsehoods lie at the very heart of our culture. Learning to lie is a natural stage in childhood development. Some kids become sophisticated liars as they age. Some grow up to believe that lies make the world a better place. As Katherine Dunn has said, "The truth is always an insult or a joke; lies are generally tastier. The nature of lies is to please whereas truth has no concern for anyone's comfort." Finally, many grow to find self-deception more comforting than self-knowledge.

Why Lies?

It is believed that deception and self-deception have played an essential role in evolution. The need to deceive has been a part of survival since the earliest beginnings. We also need to learn about ourselves and the human condition. It is an essential element of human relationships, including the relationship with self. It is believed that humanity has utilized self-deception as a survival mechanism and discovered that our capacity for dishonesty is as fundamental to us as our need to trust others. According to Y. Bhattacharjee, "Being deceitful is woven into our very fabric, so much so that it would be truthful to say that to lie is human."

White Lies

Minor lies are known as white lies. A white lie is an unimportant lie, especially one uttered in the interests of tact or politeness, one told to spare feelings or from politeness. They are lies with good intentions. They are lies told with the intent of sparing someone's feelings. They are lies about something trivial, one for which there will be few consequences if caught. They are lies to prevent an argument or bad feelings about something generally meaningless. The question is not when we tell white lies, but why. "The 15 Most Common White Lies and Why" by Marc Chernoff points out that some white lies save relationships, some ease a hectic situation, and others buy us time. The list could go on forever. Stretching the truth is a natural component of human instinct, because it is the easy way out. We all do it, so there is no reason to deny it.

What Is the Limit?

At the personal level, as long as we are not harming or hurting others, breaking the law, or doing immoral things, "innocent" white lies may make life a little more pleasant. They can absorb potential friction among our varying personalities and vacillating moods as we nudge into one another in our daily routines. Sometimes, white lies cushion us from ourselves. We just need to be careful not to harm or destroy the trust of others. As noted by Al David, "Most lies have the power to tarnish a thousand truths."

Finally, lying has shown signs of being detrimental to health, as it takes so much energy out of the liars. After all, as pointed out by Abraham Lincoln, "No man has a good enough memory to be a successful liar."

At the community level, lies can have severe consequences. Societies work under the premise that those who represent them speak with truth and integrity and consider these as prerequisites of having such responsibilities and privileges. Leaders and the media reporting lies need the trust and loyalty of their followers, readers, or listeners. Consistent and intentional utterance of untruths removes all integrity from what should be the bastion of truth and justice. Lies tear the very fabric that holds people and communities together and threaten the health of democracies.

References

Chernoff, Marc. "The 15 Most Common White Lies and Why." http://www.marcandangel.com/category/humor/.

Road Rage

Recent statistics reveal that road rage is far more dangerous than we may think. According to *Wikipedia,* road rage is aggressive or angry behavior exhibited by a driver including rude and offensive gestures, verbal insults, physical threats, or dangerous driving methods targeted toward another driver or a pedestrian in an effort to intimidate or release frustration. With this description, most people who drive regularly encounter "road rage" in one form or another. The National Highway Traffic Safety Administration (NHTSA) defines road rage as when a driver commits moving traffic offenses so as to endanger other persons or property; an assault with a motor vehicle or other dangerous weapon by the operator or passenger of one motor vehicle on the operator or passengers of another motor vehicle. Road rage can lead to altercations, assaults, and collisions that result in serious physical injuries or even death. The NHTSA makes a clear distinction between road rage and aggressive driving, where the former is a criminal charge and the latter a traffic offense. This definition of road rage places the blame on the driver.

According to a study conducted by AAA, every year, there are more than 1,200 incidents of road rage in the United States, some of which end in serious injuries or even fatalities. These rates have risen yearly throughout the six years of the study. Also, studies have shown that individuals with road rage are predominantly young (thirty-three years old on average), and 96.6 percent of them are male. There are many shocking examples of road rage. For example, in Germany, a gun-wielding truck driver was arrested in 2013 and accused of firing at more than 762 vehicles, an exceptional case of road rage. According to authorities, the Autobahn sniper was motivated by "annoyance and frustration with traffic."

The AAA has reported that student drivers and driving instructors are becoming targets of road rage at an increasing rate. According to one study presented in en.wikipedia.org, people who customize their cars with stickers and other adornments are more prone to road rage. Road rage is not an official mental disorder recognized in the Diagnostic and Statistical Manual of Mental Disorders (DSM). However, according to an article published by the Associated Press in June 2006, the behaviors typically associated with road rage can be the result of a disorder known as intermittent explosive disorder. This conclusion was drawn from surveys of some 9,200 adults in the United States between 2001 and 2003 and was funded by the National Institute of Mental Health.

Road Rage Questionnaire

Road rage could happen to anybody. If you wonder if it could happen to you, first, ask yourself the following questions presented in www.melhimesinsurance.com:

- Do I regularly drive over the speed limit?
- Do I try to "beat" red lights because I am in a hurry?
- Do I tailgate or flash my headlights at a driver in front of me that I believe is driving too slowly?
- Do I honk the horn frequently?
- Do I ever use obscene gestures or otherwise communicate angrily at other drivers?

If you answered yes to any of these questions, it is possible you are susceptible to road rage. If you answered no to the questions above, you could be causing others to lash out with road rage.

Also ask yourself the following questions (www.safemotorist.com):

- Do I frequently use my phone while driving or otherwise drive while distracted?
- Do I keep my high beams on, regardless of oncoming traffic?
- Do I switch lanes or make turns without using my turn signal?
- Do I fail to check my blind spot before switching lanes to make sure I am not cutting someone off?
- If you answered yes to any of these questions, you may be contributing to causing road rage in others.

Some Statistics

The following statistics compiled from the NHTSA and the Auto Vantage auto club show that aggressive driving and road rage are causing serious problems on our roads.

- 66 percent of traffic fatalities are caused by aggressive driving.
- 37 percent of aggressive driving incidents involve a firearm.

- Males under the age of nineteen are the most likely to exhibit road rage.
- Half of the drivers who are on the receiving end of an aggressive behavior, such as horn honking, a rude gesture, or tailgating, admit to responding with aggressive behavior themselves.
- Over a seven-year period, 218 murders and 12,610 injuries are attributed to road rage.
- 2 percent of drivers admit to trying to run an aggressor off the road!

CHAPTER 3

Medical Risks

The art of medicine consists of amusing the
patient while nature cures the disease.
–Brad Pitt

IN CARRYING OUT most medical procedures, the benefit is likely to exceed the risk by a sufficient margin to make detailed quantification of the risk seem unnecessary. In other instances, however, it is important to estimate the likely level or risk involved and to put it against the expected benefit to determine whether, or when, the use of the proposed procedure is justified. Such risks can commonly be estimated, at least in terms of some limited criterion such as the probability of death or improvement in the quality of life attributable to the procedure. When so quantified, the risk/benefit ratio proves to range widely for different conventional methods of diagnosis and treatment.

Such variations are illustrated in several articles in an examination of the number of deaths during a ten-year period that could be regarded as being due to different medicaments. One article compared these figures with number of prescriptions issued annually by general practitioners, estimating the number of deaths per million prescriptions that might be due to the use of different drugs (excluding known instances of over dosage). Of over two hundred types of drug or preparation for which prescription rates were obtainable, twenty had mortality rates of one or more deaths per million prescriptions. In one of these, the estimated rate was about 150, while the rates for the remainder ranged from one to eighteen deaths per million prescriptions.

Sometimes, the same treatment will have differing levels of risk according to the individuals concerned. Thus, the frequency with which death may result from the prophylactic vaccination of healthy people varies with a number of factors, such as the type of vaccination used and the age at which it is performed. Overall, a low fatality risk in the order of 1 in 10^6 is indicated by the average number of 3.3 deaths per year between 1967 and 1976 in England and Wales attributed to the effects of vaccinations.

During this period, an average of 3.7×10^6 vaccinations were carried out each year, against the eight types of infectious disease for which vaccinations were predominantly made.

Risk estimates, both of death and of any other serious side effects, for different types of vaccination, are important. They enable risk benefit assessments to be made as to whether, at any given time, the risk of vaccination exceeds the benefits obtained in the prevention of the disease. A decision not to vaccinate, however, requires as good an estimate of the benefit as of the risk. Hence, assessments are needed of the frequency of mortality or other effects of the disease without vaccination, as against its frequency and effects despite vaccination. The occasional fatal effects of smallpox vaccination illustrate the risk/benefit imbalance in any continued vaccination campaign, following the elimination of this disease. The actions of the World Health Organization over smallpox is one of the most cost-effective major risk reduction operations ever staged. The total cost of smallpox eradication is estimated by WHO to have been about US$300 million and to be saving an estimated two million lives per year. Whatever cash value might be assigned to a human life, it must certainly exceed the sum of $20 implied by 12 percent annual interest on the capital sum expended.

In 1982, there was considerable debate in the UK concerning the efficacy of whooping cough vaccine. In 1951, there were 169,000 cases of whooping cough, with 453 deaths. Vaccination was introduced in the 1950s with a determined national campaign in 1958. In 1959, there were 33,000 cases, with 25 deaths. By 1970, there were 16,500 cases, with 15 deaths. The public then became scared by the alleged dangers of brain damage from the vaccination, put at one in three hundred thousand injections. The immunization rate fell, and an epidemic occurred in 1979 with 100,000 cases and 30 deaths. In 1982, only about half the number of young children were vaccinated, and an epidemic was building up, which promised to be worse than that of 1979. Given annual births of around 650,000, there appears to be a direct trade-off to be made between the risk of epidemics with a substantial number of deaths and the risk of two or three brain-damaged children per annum from the vaccination, which virtually eliminates deaths from the disease.

Medical risks change over time as a consequence of research and better training. This is illustrated by deaths attributable to the anesthetic used in major surgical operations, which are linked to the length of operation, the state of health of the patient, and other factors.

Medical Science Is Not an Exact Science

Most patients expect doctors to know what their problem is and how to fix it. However, people who deal with medical-related issues point out that it is impossible to figure out the cause of every health problem even for experts and specialists.

A new study estimates that in the United States, more than a quarter million deaths occur per year because of medical errors. Medical error is defined as "an act of omission or commission in planning or execution that contributes or could contribute to an unintended result." This makes medical errors the third leading cause of death, only after heart disease and cancer. Even though this is alarming and concerning, it highlights a much larger problem, as many medical errors are not lethal.

Unlike physics or chemistry, medicine is not a pure science. Unlike mathematics, it is not an exact science. Even results obtained from sophisticated tools could have considerable variation. One pathologist may opine about a particular case as malignant, which may not be corroborated if some other colleague examines it. Additionally, scientific truths are not true for all times. In fact, today's truth may be tomorrow's folly. The half-life of truth in medicine is short. There is a saying, "Half of what is true today will be proven incorrect in the next five years and unfortunately, we do not know which half that is going to be."

For example, lab test failures contribute to delayed or wrong diagnoses and unnecessary costs and care. A 2014 study estimated that in the United States, diagnostic errors happen about twelve million times per year for outpatients alone. The Institute of Medicine concluded that most people would experience at least one diagnostic error in their life. Errors related to lab tests are more common than one might think. High-precision tests are very expensive and not affordable. Most classical tests have a relatively high percent of false alarms. Each year, more than thirteen billion tests are performed in over 250,000 certified clinical laboratories here in the United States. These include tests for genetic disorders, lead poisoning, and diabetes, and the results routinely guide diagnostic and therapeutic decisions. Despite its ubiquity, we frequently experience misperceptions that diagnostic laboratory tests are without a doubt correct.

Everyone should know that whether because of misuse or a failure mode, all lab tests have limitations. Some of the most common reasons include mistakes in ordering lab tests at the right time–and problems with the accuracy, availability, and interpretation of their results.

Here are few more sources of uncertainties: No two people respond to diseases or drugs in precisely the same way. Our body and mind are very complex, and our knowledge about their connections is very limited. Not much is known about the reaction of one's body to certain chemicals in drugs and the interaction of drugs one takes. As a result, medical field concentrates mostly on curing the symptoms, as integrated medicine is not yet well developed. Doctors have limited time, large number of patients, and lots of paperwork. In addition, there is lot to learn as the management of diseases, even diagnostic methods and ideas on causation of a particular disease, change with passage of time.

To account for the uncertainties, the medical profession has had to become adept at estimating and interpreting probability. In fact, despite doctors' best efforts, almost all the advices medical field give to the public are solely based on an expert's assessment of the probabilities involved. Depending on the situation, these probabilities are arrived at either empirically or by educated guesses. The information most doctors receive is typically determined empirically. Take, for instance, clinical trials used to determine the side effects. Here, one needs to account for uncertainties related to missing data created by subjects who do not complete their logs, who do not follow it properly, or who drop out of the study.

Similar methods are used to determine the probability of a patient experiencing a given symptom. Even once a disease has been accurately diagnosed, exactly how it will manifest itself is difficult to predict. One example of this involves the progression of HIV in infected persons. HIV can lead to cognitive impairment in its victims, referred to as HIV dementia, but this condition occurs in fewer than half of those infected. So how do doctors know who will get it? The answer is they do not know it for sure and therefore use empirical data gathering.

In sum, doctors, in all situations, weigh the probabilities when providing care to their patients. They look at not only the probability of the disease occurring to a typical person but also the probability that a certain patient will get the disease, will experience given symptoms, or will suffer from certain side effects. This means that not only is medical care always going to be slightly unreliable but also the preconceptions of a doctor will always bias the care received by the patient. In conclusion, no matter what, most patients are at the mercy of the probability and odds.

Medical Errors: The Third-Leading Cause of Death in America

This section is about a fact that is hard to believe. It includes some disturbing findings.

- According to a recent publication in the *Journal of Patient Safety*, as many as 440,000 people die each year from medical errors.
- Only heart disease and cancer kill more Americans than preventable medical errors in hospitals, a Senate panel was told on July 17, 2015.
- The United States has the most expensive health care in the world. We spend more on health care than the next ten-biggest spenders combined: Japan, Germany, France, China, United Kingdom, Italy, Canada, Brazil, Spain, and Australia. If our health-care system were a country, it would be the sixth-largest economy on the entire planet. Still, compared to the rest of the world, our health care is ranked about average—that is, our high spending is not buying us particularly safe care, said Dr. Ashish Jha of the Harvard School of Public Health.

- According to some experts, the two significant causes of such a mediocre performance are overreliance on technology and a poorly developed primary care infrastructure. We are second only to Japan in the availability of technological procedures such as MRIs and CAT scans, but unlike our expectation, this has not translated into a higher standard of care. We have one of the best systems for treating acute surgical emergencies, but our system is an unmitigated failure at treating chronic illnesses. It seems that the conventional medicine, with its focus on diagnostic tests, drugs, and surgical interventions for most ills, harms an unexpectedly large number of patients. The lethality of such system is in part due to side effects, whether "expected" or not, but preventable errors also account for an absolutely staggering number of deaths. According to a study published in the *Journal of Patient Safety*, the problem may also be linked to the "cascade effect," where diagnostic procedures lead to more treatment, more symptoms, and, hence, more complications and deaths.

- Some managers think that it is our tort law that adds to the staggering cost of medical care. But according to Tom Baker, a professor of law and health sciences at the University of Pennsylvania School of Law and author of *The Medical Malpractice Myth*, making the legal system less receptive to medical malpractice lawsuits will not significantly affect the costs of medical care. Others think that the problem is partly due to a large number of unnecessary medical and surgical procedures (around 7.5 million annually) and hospitalization (around 8.9 million annually).

Remarks

1. Most studies do not blame the physicians for all that is happening. In fact, some argue that, in many ways, physicians are just as victimized by the deficiencies of

the health-care system as patients and consumers are. With increased patient loads and mandated time limits for patient visits set by HMOs, as well as the required paperwork, most doctors are doing the best they can do to survive.

2. Some find the outcome not at all surprising. For example, homoeopathic practitioners believe that the medical science have gone too far in the wrong direction, and the more they go down that path, the worst things are sure to happen.

3. Some do not see a need for change as they see no cause/effect relationship in published reports, only conjecture based on statistics. They argue, for example, that if ten thousand people die while undergoing medical treatment, it cannot be taken to imply that these people died because of undergoing medical treatment. It is just as likely or even more likely that these same people would have died if there had not been any medical treatment.

Is There an Alternative?

Most experts agree that there isn't any easy fix. Some argue that any profession is, and always will be, within the confines of some for-profit system. To suggest that there is a solution outside of this reality is naïve. Within the medical field, someone should always profit. The question is how to put this in line with patients' interests or benefits.

Some suggest minimizing interactions with the conventional system, which, at least in the case of chronic disease, has little to offer. Some blame the conventional strategies as they often target the symptoms and not the underlying cause of the disease. They suggest a gradual transition to so-called integrated medicine based on mind-body connection. Integrated medicine combines the most scientifically validated and least harmful therapies from both high-tech and holistic medical practices. It seems that some doctors and patients alike are bonding with this philosophy and its whole-person approach—designed to treat the person, not just the disease. They think that therapies that take advantage of the subtle interactions between a person's state of mind and basic physiological functions in their body are a reasonable alternative. This approach include the mind-body medicine, which uses relaxation techniques and the power of thoughts and emotions.

Risk and Diagnostic Tests

Diagnostic tests are frequently used to detect presence or absence of a disease. For example, mammogram is used to detect breast cancer in women. Prostate-specific antigen (PSA) test is used to detect prostate cancer in men. Ordinarily, *positive* test result indicates presence and *negative* test result *indicates* absence of the disease. But how often do test results match with reality? This is usually assessed by two numbers known as *sensitivity* and *specificity* of the test. Sensitivity is the probability (percentage) of a positive test result when disease is present, and specificity is the probability (percentage) of a negative test result when disease is absent. The higher the sensitivity and specificity, the better the diagnostic test.

Despite their importance, sensitivity and specificity are not of main concern to most patients. Instead, informed patients want to know what percentages of people who test positive actually have the disease—that is, how often positive test results indicate disease. In some cases, the answer to this question surprises both doctors and the patients. In fact, when relatively few people have the disease, the probability that patient with positive test result actually has the disease can be surprisingly low even for tests with a high sensitivity and specificity. For example, consider the breast cancer in women. It has been estimated that of women who get mammograms at any given time, only 1 percent truly have breast cancer. (Note: this number should not be confused with likelihood of breast cancer in women's lifetime). For mammograms, typical values reported are 0.86 for sensitivity and 0.88 for specificity. For a positive test result from a mammogram with these specifications, the probability that the woman truly has breast cancer is only $0.077 < 8$ percent. Surprised?

How can this probability be so low, given a relatively good sensitivity and specificity for the test? Consider a random sample of one hundred women selected for screening. Here, one woman is expected to have breast cancer, 1 percent of the sample. For the woman with breast cancer, the chance of detecting it is 0.86. So we would expect the one woman with breast cancer to have a positive test result. For a woman without breast cancer, the chance of detecting it is 0.88. So for this group of ninety-nine women, we would expect about $0.88 \times 99 = 87$ negative results and $0.12 \times 99 = 12$ positive (false positive) results. This shows that of the thirteen women with a positive test result, the proportion $1/13 = 0.08$ actually have breast cancer.

Next, consider the prostate-specific antigen (PSA) blood test for prostate cancer in men. Autopsy studies suggest that about half of all men over fifty have cancerous cells in their prostate, but only about 2.4 percent die of prostate cancer. The PSA test has high false positive and false negative rates (e.g., about 50 percent of the men with high PSA readings do not have prostate cancer); when a person tests positive, there are other tests to do, which are often inconclusive. Given that cancerous prostate cells might never pose a health threat, a decision to whether PSA tests should be given routinely or not is not an easy one. The current recommendation by the American Cancer Society is that men over the age of fifty should have an annual PSA test, along with a digital rectal exam, although a statistical analysis published in the1994 issue of the *Journal of the American Medical Association* suggests that the benefits of such screening are marginal at best.

Drinking: The Fourth-Leading Cause of Death in the United States

Drinking has been an accepted part of Western culture for generations. It is associated with everything from weddings to wakes. Studies show that moderate consumption of alcohol decreases the risk for heart disease, ischemic stroke, and diabetes. However, its excessive misuse creates many problems and takes an extremely heavy toll on families and society.

Alcohol

Alcohol, a semiluxury beverage, is also a drug that acts on the central nervous system. Although it had always been regarded either as a bad habit or a sin, the realization that alcoholism is a disease only came at the turn of the century. Today, alcoholism is one of the world's greatest health problems. Apart from environmental, sociocultural, and psychological influences, recent epidemiological studies have provided clear evidence of a genetically conditioned predisposition toward alcoholism. Moderate consumption of alcohol decreases the risk for heart disease, ischemic stroke, and diabetes. But its excessive use or misuse creates many problems and takes an extremely heavy toll on families and society. Despite all the studies, it is extremely difficult to draw the line between genetic influences on the one hand, and social, cultural, and environmental factors such as religious upbringing, the price, and availability in the other.

In recent years, there has been a considerable amount of scientific research on the effects of alcohol on memory, reflexes, coordination, and depth perception as well as a host of other cognitive and psychomotor processes. Formulation of such research questions requires quantification of the amount of alcohol in the blood. This has led to the introduction of blood alcohol levels as a percentage of alcohol per volume of blood.

A blood alcohol level (BAC) of 0.10 is defined as one gram per kilogram of blood–meaning, alcohol is 0.10 percent, or 1/1,000, of the blood. For example, a person weighing 150 pounds with 34 grams of alcohol in the body, the amount of alcohol from two and a half beers would have a 0.08 BAC equal to 0.08. The same person with 210 grams of alcohol in the body would have a 0.50 BAC.

Quantification of blood alcohol level helps the investigation of many other questions and opens the door for introduction of other concepts found in elementary mathematics and related subjects. For example, it has been demonstrated that the probability of being involved in an alcohol-related accident increases as the BAC increases. Estimates show that an individual with a BAC between 0.10 and 0.14 is 48 times more likely to be involved in a motor vehicle accident than an individual who has not been drinking. This can be used in discussion of probability and its definitions.

The BAC of an individual is determined by three primary factors: body weight, amount of alcohol consumed, and amount of elapsed time from first drink until a breath and/or blood sample is taken. As an example, a 180-pound man who consumes seven drinks over a three-hour period of time will have a BAC of around 0.10. In contrast, a 110-pound man who consumes the same seven drinks over three hours will have a BAC of around 0.20. One drink, which is defined as one ounce of liquor, five ounces of wine, or one 12-ounce beer, consists of approximately 14 grams of alcohol. Note that all these information and data can be used for developing lesson plans. The following is a specific example of alcohol-related information useful for teaching functions, an important concept in elementary mathematics.

After a drink, the alcohol begins to enter the bloodstream almost immediately, and the blood-alcohol level rises rapidly. Once the person stops drinking, her natural metabolic processes slowly eliminate the alcohol, and blood-alcohol level begins to fall. So how long must one wait after drinking for safe driving?

Experiments show that after drinking six beers, four glasses of wine, or four shots of liquor, the blood-alcohol level typically rises to 0.11 gram per deciliter of blood. Thereafter, alcohol is eliminated at the rate of 0.02 gram per hour. The graph below shows the resulting blood-alcohol level over time, first, rising rapidly and then falling linearly. By calculating where the graph falls below the legal limit, we can find how long one must wait. This relationship can be described by an important mathematical tool known as function. Functions have different types and can be classified according to their properties. One important classification is linear versus nonlinear. To introduce these, one can refer to the fact that many substances are eliminated from the bloodstream nonlinearly, but alcohol is eliminated linearly.

Global Alcohol Statistics

Alcohol, a semiluxury beverage, is also a drug that acts on the central nervous system. Intemperate indulgence had always been regarded either as a bad habit or a sin, but the realization that alcoholism is a disease only came at the turn of the century. According to the World Health Organization, alcohol misuse causes 3.3 million deaths annually and contributes to more than two hundred diseases and injuries. Globally, it is the fifth-leading risk factor for premature death and disability among those aged fifteen to forty-nine and the first risk factor for those twenty to thirty-nine years old.

USA Alcohol Statistics

In the United States, alcohol misuse cost approximately $250 billion annually. It is responsible for approximately 88,000 deaths (62,000 men and 26,000 women) per year, making it the fourth-leading preventable cause of death. In 2013, of the 72,559 liver disease deaths in age group twelve and older, 45.8 percent were alcohol related. It was also the primary cause of almost one in three liver transplants. Drinking alcohol increases the risk of cancers of the mouth, esophagus, pharynx, larynx, liver, and breast. The article "Statistics on Alcoholics" contains additional data on alcohol, such as some of the following:

- Alcohol is the number-one drug problem in America.
- There are more than twelve million alcoholics in the United States.
- In the United States, eighteen thousand people are killed in an alcohol-related car accident every year.
- In 2000, nearly seven million ages twelve to twenty were binge drinkers. Binge drinking is defined as having five or more alcoholic drinks on one occasion for men, or four or more for women.
- 75 percent of all high school seniors report being drunk at least once.
- Young people who begin drinking before their fifteen birthday are four times more likely to become alcoholics than those who do not begin drinking until the age of twenty-one.
- Unlike what is expected, people with a higher education are more likely to drink.
- People with higher income are more likely to drink.
- Excessive drinking is responsible for more than 4,300 deaths among underage youth each year.
- People ages twelve to twenty drink 11 percent of all alcohol consumed.

The statistics on alcoholics listed above do not include victims who may not even drink at all. Alcohol contributes to 73 percent of all felonies, 73 percent of child beatings, 41 percent of rapes, 81 percent of wife beatings, 72 percent of stabbings, and 83 percent of homicides. Finally, while there are twelve million alcoholics, an estimated forty million to fifty million people, such as family members, suffer the consequences of alcoholism.

College Students

- According to the 2015 report, 58.0 percent of full-time students, ages eighteen to twenty-two drank alcohol in the past month compared to 48.2 percent of other persons of the same age.
- 37.9 percent reported binge drinking in the past month compared to 32.6 percent of other persons of the same age.
- 1,825 students between the ages of eighteen and twenty-four die from alcohol-related unintentional injuries, including motor-vehicle crashes.
- 696,000 were assaulted by other students who has been drinking.
- 97,000 reported experiencing alcohol-related sexual assault or date rape.
- Roughly 20 percent meet the criteria for alcohol use disorder.
- About 25 percent reported academic consequences for drinking, including missing class, falling behind in class, doing poorly on exams or papers, and receiving lower grades overall.

Everything Has a Price

The excessive use or abuse of alcohol creates many problems and takes an extremely heavy toll on families and society. It is an example of the price societies pay for what is referred to as freedom of choice. The main drawback of such freedom is the price others have to pay for the choices we make. Alcoholism is one of the world's greatest health problems. Apart from environmental, sociocultural, and psychological influences, recent epidemiological studies have provided clear evidence of a genetically conditioned predisposition toward alcoholism.

Is There Any Safe Level of Alcohol?

While some medical studies–and a great deal of media attention–have focused on possible health benefits of drinking alcohol in moderation, a large new report published in a reputable journal the *Lancet* in August of 2018 warns that the harms of alcohol greatly outweigh any potential beneficial effects. The study looked at data on twenty-eight million people worldwide and determined that, considering the risks, there is "no safe level of alcohol." "The conclusions of the study are clear and unambiguous: alcohol is a colossal global health issue and small reductions in health-related harms at low levels of alcohol intake are outweighed by the increased risk of other health-related harms, including cancer."

References

"Statistics on Alcoholics." https://www.learn-about-alcoholism. com/statistics-on-alcoholics.html.

Obesity: An Epidemic

Obesity is now approaching epidemic proportions globally. Today, more than 1.8 billion adults are overweight and 320 million are obese. A recent comprehensive study including 19.2 million adults in 186 countries revealed that in the last forty years, the average global body mass index has risen by the equivalent of 3.31 pounds per person, per decade. High-income English-speaking countries account for some of the biggest rises. They also account for more than a quarter of severely obese people in the world. As expected, the United States together with China have the most obese people in the world, with the United States having the most severely obese people of any country. In fact, based on the data collected last September, in United States, the obesity rate exceeds 35 percent in three states, 30 percent in twenty-two states, 25 percent in forty-five states, and 20 percent in every state. The annual health costs of obesity are now around 21 percent of our total medical costs. This is mainly because there is a link between obesity and more than sixty chronic diseases, including diabetes and cardiovascular diseases. The only bit of good news is that the trend of increasing obesity in United States has slowed since the year 2000.

As for what is driving America's chronic weight problem, there are no definite answers. Many theories are out there, including contribution of genetics, age, and lifestyle. So far, the preponderance of evidence points to the two causes most of us already suspect: too much food and too little exercise. The role of food and diet is obviously major, but it's also complex. We receive mixed messages when it comes to what to eat and how much. As a result, we have developed a culture of looking for fast food, and at the same time, fast weight-loss options. We spend more time at work and less time in our homes and kitchens than our parents. We consume 31 percent more calories now than forty years ago. We walk less than people in any other industrialized country. This is even true for our youngsters. Presently, less than 15 percent of our school-aged children walk or bike to school, compared to 48 percent that did in 1969. So, as expected, 13.9 percent of our high school students are obese, and an additional 16.0 percent are overweight.

So, considering several negative effects of obesity, we all should be concerned not only for our own heath but also for the sake of our children, grandchildren, as well as our country and the world. Research shows that many problems related to obesity can be avoided by simple changes in lifestyle. This includes staying active and eating healthier food and smaller portions of it.

CHAPTER 4

Environmental Risks

Let us not risk the Earth; after all, that
is what we have in common.

I N RECENT YEARS, environmental risk has become a subject of great importance and concern. This chapter is devoted to the risk of natural disasters and the related issues.

Natural Disasters

Natural disasters are defined as a rapid instantaneous or profound impact of the natural environment upon the socioeconomic system. They do not recognize gender, race, religion, etc., and claim lives of many innocents, especially those who are most vulnerable. Although natural disasters occur frequently and cause a great deal of destruction and suffering, they do not receive the attention they deserve unless they are close to home. Natural disasters are hardly a subject of lobbying. One look at the budget allocated to events affecting celebrities versus those effecting underrepresented people makes it clear how such decisions are made.

All of this happens despite the fact that, in recent years, the world has witnessed many catastrophic natural disasters. The residential and commercial development along the coastlines and areas that are prone to disasters suggest that future losses will only grow—a trend that emphasizes, as never before, the need for further research, investigation, and education. Recently, a team of scientists has found that the increase in economic damages from, for example, hurricanes in the United States, is due to greater population, infrastructure, and wealth along U.S. coastlines—not due to a spike in the number or intensity of hurricanes. According to their investigation, the economic hurricane damage in the United States has doubled every ten to fifteen years.

Hazard assessment and risk analysis of natural disasters require detailed analysis of the available information. The raw material of such investigations are the past data, which are often incomplete and insufficient. Because of the nature of the problem, statistical methods have been increasingly used in hazard assessment and risk analysis of disasters such as earthquakes, high-speed winds, and floods. Although most of the classical mathematical and statistical theories have been utilized for this purpose, there are still many important areas that have not received enough attention. Few years ago, I tried to look for more applications of both mathematics and probability and put together my findings in a book titled *Statistical Methods for Earthquake Hazard Assessment and Risk Analysis*, published by Nova Science Publishers Inc., Commack, New York. The book contains description of some new methods and theories and discussion of their relevance to hazard assessment of natural disasters in general and earthquakes in particular.

Hazard assessment may be defined as the process of estimating the probability that certain performance variates at a site of interest exceed relevant critical levels within a specified time period as a result of nearby events. The process has a multistage structure with an assessment requiring critical investigation of several pieces. The choice, construction, and estimation of the chosen performance or strength variate (mark) are crucial to the hazard assessment. In this book, several marks and their analysis and modeling are discussed, and some examples of their use in earthquake hazard assessment and risk analysis are presented. Some of these methods are based on frequency-magnitude relation developed by Gutenberg and Richter and its modification and extension involving some new models. Others make use of extreme value theory, threshold theory, and theory of records discussed in this book. When modeling the seismic record, particular emphasis is placed on models whose parameters have a physical interpretation so that validation other than statistical goodness of fit is possible. Also, several models are introduced for stress and strength of systems or structures subject to earthquakes, and these, together with stochastic models for occurrences of earthquakes, are used for risk analysis and reliability calculations. We note that each of the methods and procedures presented have their own strength and weakness. Although discussions are carried out using earthquakes, most methods presented can be applied to winds, floods, and other similar natural disasters directly or with some minor modifications.

Except for introductory material, chapters of the book are presented in a format independent from each other whenever possible. This is done at the expense of being partially repetitive. In doing so, the goal was to make the book more accessible to the readers who seek to study and apply a particular method. Also, to facilitate this, relevant theories are presented in detail in the chapters where methods are discussed.

The book is useful to statisticians of all levels and earth scientists as well as student and professional actuaries, environmental scientists, civil engineers, earthquake engineers, and reliability engineers. It is also useful to practitioners and experts in risk analysis and people who are interested in natural disasters and their management.

Consequences of Natural Disaster

Natural disasters are a potential threat to the world community, and to reduce their impact, complete international cooperation is essential. In the last two decades alone, natural disasters have claimed more than three million lives and disrupted the lives of over a billion people. Official investigation shows that the economic damage from natural disasters has quadrupled in the last thirty years and that last decade's costs were at least double those of the decade before.

Disasters degrade the social and economic conditions of those most vulnerable in society, resulting in deteriorating health and a decline in education and other social services. For some nations, the political economy of disasters results in negative effects on balance of payments, increased food import needs, and opportunity costs caused by the diversion of development resources to relief activities. Ensuing economic consequences also leads to a degradation of the social structures that uphold human rights, resulting in a further marginalization of those already discriminated against because of class, gender, race, or religion. It is, therefore, important to situate disasters in a greater perspective and view relief, disaster preparedness, and development as part of the same continuum. The world can no longer afford to take a pay-as-you-go approach to natural disasters and the consequences.

The United Nations has designated October 11 as the International Day for Natural Disaster Reduction. This day reminds us of great human suffering and losses. The experience of recent years shows that improved public education is essential to the success of natural disaster reduction, which aims to reduce, through appropriate action, loss of life and property damage. While activity at the local or national level is necessary, the world needs active and informed citizens furthering safe societies on a global scale as well. Such a global citizenry may be promoted through school curricula, as well as within the overall precollege curriculum, social studies, and especially geographical education. Within this curriculum, the emphasis must be on the study of disaster rather than on a list of behaviors to be adopted in the event of impending disaster.

It is important to note that disasters result from people's vulnerability to natural events and their inability to cope with these natural forces. Moreover, vulnerability itself is a result of social, cultural, and economic processes. Thus, while schools can and should promote appropriate behavior in the face of potential hazards, they also have a responsibility to help young people understand the physical and human systems that, in certain combinations, lead to disasters.

Education in a larger scale is, of course, the responsibility of the media. After all, natural disasters do not respect borders, race, or religion. They are really everybody's problem and nobody's fault, and those usually most affected are the socially disadvantaged groups who are least equipped to cope with them. Making disaster reduction a priority in public policy is essential if we want a safer, healthier, and more productive world in the twenty-first century.

Global Warming

Global warming is the rise in average temperature of Earth's atmosphere and oceans. It is a matter that should be taken extremely serious by all of Earth's inhabitants. Since 1980, Earth's average temperature is said to have increased by approximately 0.5 degrees Celsius. Scientists are also predicting the temperature to continue increasing by another 2.4 to 6.4 degrees Celsius by the next century. In short, this may not mean a lot, but if rates continue the way they are now, the future of the human race could be in loads of trouble.

Scientists believe there are a number of reasons that could be causing this slow increase of Earth's temperature. With 90 percent certainty, scientists think the main causes are the increase of greenhouse gases and burning of fossil fuels. It can also be observed by watching the size of the ice around the North Pole. Over the past ten years, the ice has been slowly shrinking with it being at its smallest distance ever recorded.

There is a lot of controversy over whether or not the global warming is actually happening, or if it even matters. With the extremely high amounts of CO_2 in the atmosphere and the constant deforestation happening, many believe that nothing is going to get any better.

Statement of the Problem

In recent years, much discussion is devoted to global warming and the related issues such as the contribution of greenhouse gas emissions from human activities. Because of the complexity of the problem and uncertainty involved, some studies have reported conflicting results leading to more questions than answers. A primary source of the problem, in my view, relates to how data were compiled and, more importantly, how the statistical analysis has been carried out. For example, an analysis of the minimum/maximum temperature data in the Antarctic Peninsula shows a significant increase in the maximum temperatures but a very little change in the minimum temperatures. So analyzing minimum/maximum temperatures could lead to a different conclusion. As a result, studies that included ozone levels into the model might have found it to be a significant/nonsignificant factor.

This project plans to examine data collection methods and statistical procedures applied in global warming studies. I anticipate facing questions and problems that have not been addressed before. Through this investigation, I hope to show that this and interpretation of results are partly responsible for differences one finds in publisher reports. I am planning to utilize an approach different from those one finds in the literature. If successful, the outcome could partially explain why, despite many signs of global warming, global dimming, and changes to the climate, there are still some conflicting results and views about what is taking place. The findings could also serve as educational materials for teachers and instructors.

Significance of the Problem

Statistics is used in most research and investigations. Statistical methods are intended to aid the interpretation of data that are subject to appreciable haphazard variable. The methods are eclectic, and, consequently, it is often difficult to decide which of several ways of analyzing data is most appropriate. In fact, there is more to the correct use of statistics than knowledge of classical statistical techniques and use of statistical software. Investigations of published research in some critical disciplines have unveiled many unintentional misuses of statistical methods as well as incorrect application or interpretation of the results. Clearly, decisions based on incorrect statistical analysis could lead to serious consequences. For example, methods applied to a complete data set are not directly applicable to data sets with missing observations. This is the case for global warming problem since it is not possible to make regular measurements during the Antarctic winter when station is in darkness. Additionally, the available data is from only one station (Antarctic). Other factors making the inference difficult are the following:

- It is known that Antarctic air is warming faster than the rest of the world.
- The Antarctic Peninsula has experienced major warming over the last fifty years.
- It is believed that the increase in mean surface temperature at this station is mainly due to the increases in minimum temperatures.
- For the period of interest, 1951–2004, the minimum/maximum monthly temperatures are separately available only at this station.

Considering these, I think it is beneficial to examine statistical methods used to study global warming phenomenon, especially the part that involves factors from both space and anthropogenic activities. For example, consider the sunspot number that affects global temperature. It is known that sunspot number is periodic with a period of eleven to twelve years. As a result, the analysis of different segments of data could lead to different conclusions. In fact, the periodicity of sunspot numbers may have something to do with the fact that eleven of the twelve warmest years on record have occurred in the past fourteen years.

Pollution

The American Heritage Science Dictionary defines pollution as the "contamination of air, water, or soil by substances that are harmful to living organisms. Pollution can occur naturally, through volcanic eruptions, for example, or as the result of human activities, such as the spilling of oil or disposal of industrial waste." Some people believe that our planet is being destroyed at an alarming rate because of the way the human race has treated it over the past few centuries. They point out that advancement has come with a high rate of pollution. As a result, millions of people do not have access to clean drinking water, often because of industrial runoff invading bodies of water. In addition, animal species are going extinct and chronic illnesses are rising.

Air pollution has damaging effects on human cognitive ability, according to new research published in the journal *Psychological and Cognitive Science*. According to a 2018 report from the Health Effects Institute, nearly 95 percent of the world's population currently live in areas with air pollution that exceeds global air quality guidelines. Looking at data from nationwide cognitive tests in China, scores of verbal and mathematics questions for over thirty-one thousand individuals were compared with air quality data from 2010 to 2014. They found that polluted air impairs cognitive ability as people age.

According to the World Health Organization, air pollution has increased by 8 percent in the past five years. This data that was compiled from three thousand cities has produced these alarming statistic. China is no longer the country with the most polluted air. This is not due to their government taking steps to clean their air, but because other countries are becoming increasingly more polluted. Some of the most polluted countries include Bahrain, India, Iran, and Egypt. Americans make up about 5 percent of the global population but consume 25 percent of the world's natural resources. Every year, people use more of the world's natural resources than can be produced. Soon enough, there will be no fuel to burn, water to drink, or food to eat. Because of the constant pollution of our planet, there has been an exponential rise in chronic illnesses such as lung cancer. Lung cancer rates have spiked in cities, and people living in them have a 20 percent risk of developing the disease. The risk for those who live in suburban or farm areas is 6.67 percent. With pollution constantly growing and the high yearly depletion of natural resources, the future of our world looks dismal.

According to the article "Top 10 Most Polluted Countries in the World," living in one of the cleanest countries in the world ensures a happy and healthy life not just for you but for your family. Young children who need special care especially benefit from growing up in a clean environment. There are several countries around the world that are ranked as the most polluted, having dirty air that can destroy respiratory systems. The polluted air contains countless particles such as dirt, smoke, pollen, mold, and more particles that pollute the air. They can be easily inhaled and may accumulate in the respiratory system to cause many chronic diseases. In 2012, 3.7 million people died as a result of air pollution. The best indicator that is used for assessing health impacts that result from air pollution is PM2.5, and measuring the concentration of air pollution is in micrograms per cubic meter (ug/m3) of air.

Here is a quick glance at the top-ten most polluted countries in the world with the dirtiest air and the highest levels of air pollution. The countries that are presented here are ranked on the basis of the average PM2.5 pollution.

- Bahrain
- India
- United Arab Emirates
- Mongolia
- Egypt
- Iran
- Bangladesh
- Afghanistan
- Qatar
- Pakistan

The countries with the least polluted urban areas include

- Australia
- New Zealand
- Finland
- Canada
- Iceland
- Sweden

References

"Top 10 Most Polluted Countries in the World." https://www.topteny.com/top-10-most-polluted-countries-in-the-world/.

Flood Risk: Should I Buy Insurance?

If a bus ticket costs one dollar and the fine for fare-dodging is twenty-five dollars, at what point is it rational not to buy a ticket?

Since moving to Bloomsburg and learning about past destructive floods, I always wondered how residents close to the major rivers evaluate their risk associated with floods. To find out, I first talked to a few friends who live or have lived in a flood plain both before and after significant floods. What I heard was interesting, but mostly personal, especially with regard to buying flood insurance. I realized that almost everybody did their best to find an answer to the following critical question: at what point is it rational to take risk and not purchase insurance? Or at what point is it reasonable to purchase flood insurance? I also noted that some of them have realized that the key to an informed decision was to come up with a reliable estimate for the frequency of future destructive floods.

Next, I put myself in their situation and tried to approach the problem in a little more systematic way. Below is a summary of this effort. These calculations are based on thirty-eight past floods, all greater than 19.8 feet that have occurred in Bloomsburg since 1850 (water.weather.gov). They predict the expected number of future floods and the probabilities related to them:

1. During the next ten years, two floods greater than 19.8 feet are expected.
2. The probability of a flood greater than 32.75 feet (2011 flood) during the next ten, twenty, thirty, and one hundred years are, respectively, 5.81 percent, 11 percent, 15.6 percent, and 45 percent.
3. The probability of a flood greater than 33 feet at any time in the future is less than 10 percent. In other words, we

are 90 percent confident that the largest possible flood in Bloomsburg will not exceed 33 feet.

Now, is any other helpful information available? To be honest, not much. At best, only the fact that we live in an endangered area. This is because the subject is very complicated, Earth is much too chaotic, and our knowledge is so limited. As such, it is hard to develop any reliable risk prediction model for future floods. So, for now, all experts can do is to try to understand the processes better and, with the help of experts from related fields, develop guidelines for a more efficient management of natural disasters in general and floods in particular. So, until then, a sensible approach would be to follow some of my friends and do cost-benefit analysis using numbers such as the ones above, if useful, and purchase flood insurance, if affordable. And in doing so, remember that although, like any other insurance, flood insurance has a long-run negative return, one is purchasing peace of mind too.

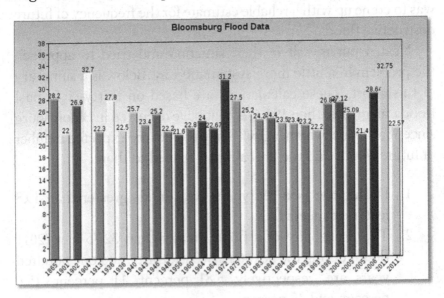

Summary

Natural disasters continue to claim the lives of many innocents, especially those who are most vulnerable. The impact of natural disasters, in terms of human and economic losses has risen in recent years, and society has become more vulnerable to natural disasters. While the number of lives lost has declined in the past twenty years, the number of people affected has risen in the past decade. As the world's population increases and areas previously almost uninhabited become more heavily settled, the propensity for natural disasters to cause damage also increases. Other than physical effects, disasters degrade the social and economic conditions of those most vulnerable, resulting in deteriorating health and a decline in education and other social services.

Considering these, it is important to situate disasters in a greater perspective and view relief, disaster preparedness, and development as part of the same continuum. The world can no longer afford to take a pay-as-you-go approach to natural disasters and their consequences. While activity at the local or national level is necessary, the world needs active and informed citizens furthering safe societies on a global scale as well. Such a global citizenry may be promoted through school curricula and within the overall precollege curriculum, social studies, and especially geographical education. Within this curriculum, the emphasis must be on the study of disaster rather than on a list of behaviors to be adopted in the event of impending disaster.

Hazardous Waste

Twenty years ago, comparatively few people were much concerned about the management of hazardous wastes. But extensive newspaper and television coverage of the subject is making people aware of dangers. A recent Roper poll showed that the two most pressing environmental concerns among Americans are used and abandoned hazardous waste sites. More than two-thirds of those surveyed thought that active hazardous waste sites were a serious problem, and 62 percent thought that abandoned sites were problematic. Politicians who monitor such surveys are likely to conclude that industry should strive for waste-free production and that more funds are needed to clean up sites. How much should we spend to do that?

Some surveys show that the public is ready to pay as much as necessary to clean up the environment. One sign of this readiness was the response by a cross-section of people asked whether they agreed or disagreed with this statement:

> Protecting the environment is so important that requirements and standards cannot be too high and continuing environmental improvements must be made regardless of cost.

In 1983, 58 percent of those surveyed agreed. When a *New York Times*/CBS News survey posed this same question in 1990, the percentage had risen to 74 percent.

Other studies continue to be made. One conducted at the University of Tennessee examines public attitudes toward how to punish those who dump wastes. A related study at the Wharton Risk and Decision Process Center seeks to measure what people think about who should pay for cleanup. If, the study asks, a company followed government hazardous waste standards but did not use state-of-the-art technology, what proportion of the cleanup cost should it pay if groundwater becomes polluted?

Experts say that the results of such studies often point to a confusion in public perception. Scientists do not agree about the nature of risks associated with hazardous waste, but even if they did and could place an exact price on each step necessary to deal with the risks, choices would still have to be made. In risk management—as in deductibles for home insurance—if we want to decrease our risk, we should expect to increase our cost. And hazardous waste sites constitute only one of many social risks

Americans face daily. So the demand by many people for a no-risk hazardous waste policy ("requirements and standards cannot be too high"; "improvements must be made regardless of cost") may be not only unreasonable but also impossible.

All of us, experts agree, will have to bear some costs for a cleaner environment. We will have to assess the risks we face and decide how dangerous each is and how much we should spend to control it. But many experts believe that American policy on the environment has become paralyzed, that it is hostage to irrational fears, that our ability to balance benefits against risks has become impaired. Our resources, they believe, are not being used efficiently. Experts generally are convinced that hazardous waste facilities pose medium to low risks. But people affirm in poll after poll that they consider such sites high-risk hazards. It is no longer possible, experts complain, to gain public acceptance to build potentially hazardous facilities, even when the benefits of those facilities far exceed the risks.

CHAPTER 5

Risk and Investment

To win without risk is to triumph without glory.
–Pierre Corneille

IN TODAY'S BUSINESS, both managers and individual investors face risk and uncertainty. This chapter discusses the risk of the security by considering its two main components: the unique risk, which can be diversified by means of portfolio, and market risk, which cannot be removed by individual action. The measure and the method used are, respectively, the sensitivity of the stock change to market change and regression analysis. Data from Delta Airline and Disney Company are used for illustration. A brief discussion concerning the recent developments on the subject is also included.

Introduction

In the context of investing, risk usually refers to the probability of loss or the dispersion of actual from expected results. In the world of business, risk has a profound implication to the investors. So knowing how to deal with risk is very important. There are two kinds of risk for a security. One is the unique risk; the other is the market risk. Here, these two kinds of risk, their measurement, diversification, and estimation are discussed.

Unique Risk and Its Diversification

Unique risk stems from the fact that many of the perils that surround an individual company are peculiar to that company and perhaps its immediate competitors. Stock market is considered risky because there is a spread of possible outcomes. A usual measure of this spread is the standard deviation of the return. Note that the bigger the standard deviation is, the riskier the security is. Fortunately, the price of different securities do not move exactly together. So we can select some kinds of securities to set up a portfolio to diversify the unique risk. In the following part, we will discuss how the portfolio works.

A portfolio is a group of securities. Suppose we have several types of stocks. We define an index involving each security's standard deviation and the correlations between that pair of securities, weighted by the proportions invested in each securities.

Here, correlation coefficient measures the degree to which the two stocks "covary." So if stocks in the portfolio are relatively independent, correlation would be zero. For a very large number of stocks, the portfolio correlation would tend to zero. Of course, this is only a theoretical assumption. In fact, for a portfolio with large numbers of stocks, correlation of any pairs of stocks cannot be zero.

Second, if we choose the stocks carefully, the correlation will be small and even a negative figure. In this case, the correlation will give us a deductible part from the average standard deviation, and hence the risk is diversified. Thus, we can conclude that the investors can choose some kinds of securities to set up a portfolio, and the portfolio can diversify the unique risk of the security.

Market Risk and Its Estimation

As discussed, the investors can eliminate unique risk by holding a well-diversified portfolio, so what is left is the market risk associated with market's wide variations. Thus, the risk of a fully diversified portfolio is the market risk.

A stock's contribution to the risk of a fully diversified portfolio depends on its sensitivity to market changes. This sensitivity is generally known as beta, denoted by the following: A security with = 1.0 has average market risk; a well-diversified portfolio of such securities has the same standard deviation as the market index. A security with = 0.5 has below-average market risk; a well-diversified portfolio of these securities tend to move half as far as the market moves and has half the market's standard deviation. In other words, for the security with = 1.0, if the market index changes 20 percent, its return (or price) would change 20 percent too. While for the security with = 0.5, if the market index changes 20 percent, its return (or price) would change 10 percent; and for the security with = 2.0, if the market index changes 20 percent, its return (or price) would change 40 percent. These relations can be shown by the slope of the line. So, to the investors, what really matters is the security's beta value and the way it can be estimated. Fortunately, the regression analysis provides an effective approach to get the answer. And the development of the computer software makes this task much easier.

Discussion

In this part, the risk of the security is discussed. For the unique risk, investors can diversify it by means of portfolio. What really matters is the market risk, which cannot be removed. To measure this risk, we should use the sensitivity of the stock change to the market change. The approach often used is based on the regression analysis using. What the method does is estimation of beta value using the available data that includes the past information. Note that although the beta value does appear to be stable, one should remember that it is just an estimation. However, despite the difficulties, it is still possible to calculate the risk if we desire to invest in the capital market.

β and Investment Risk

Numerous factors affect markets, and investing requires basic understanding of the risks involved. For example, risk of a security has two main components: unique risk and market risk.

Diversification versus Risk

Investors with a low tolerance for risk try to reduce or avoid it by actions such as diversification. Unique risk can be reduced or even eliminated by holding a well-diversified portfolio. Market risk associated with market's wide variations cannot be removed by individual action.

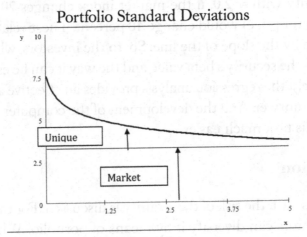

Portfolio Standard Deviations

Number of shares

As such, the risk of a fully diversified portfolio is the market risk or its sensitivity to market changes. This sensitivity is generally measured by a quantity known as beta (β).

A security with $\beta = 1.0$ has average market risk; a well-diversified portfolio of such securities has the same risk as the market index.

A security with $\beta = 0.5$ has below average market risk; a well-diversified portfolio of these securities tend to move half as far as the market moves and has half the market's risk.

For the security with $\beta = 1.0$, if the market index changes 20 percent, its return (or price) would change 20 percent too.

For the security with $\beta = 0.5$, if the market index changes 20 percent, its return (or price) would change 10 percent.

For the security with $\beta = 2.0$, if the market index changes 20 percent, its return (or price) would change 40 percent. The regression analysis provides an effective approach for finding β.

As discussed earlier, deciding on a risk definition involves trade-off between ease of estimation, forecast ability, calculation of portfolio risk, and intuition of individual investors. As such, all the measures presently used by investors have some shortcomings. All definitions of risk arise fundamentally from the probability distribution of possible returns. As such, it is complicated and full of detail. Hence, all the definitions of risk will attempt to capture in a single number the essentials of risk more fully described in the complete distribution. Each definition of risk will have at least some shortcomings because of this simplification.

Different definitions may also have shortcomings based on difficulties of accurate forecasting or their application to the risk analysis of a portfolio.

Proposed Risk Definitions

The standard deviation measure the spread of the distribution about its mean. Investors commonly refer to the standard deviation as the volatility.

If these returns were normally distributed, then two-thirds of the monthly returns would have fallen within 6.3 percent of the mean–that is, in the band between -4.7 percent and 7.9 percent. These statements are based on empirical rule. Here, we can also make statements using Chebyshev's rule in place of empirical rule. For example, at least in three-fourths of the year's fund's annual returns were in a band -24.4 percent and 62.8 percent.

Because of certain considerations, the standard deviation was Harry Markowitz's, who started such analysis definition of risk, and it has been the standard in the institutional investment community ever since. It is a very well-understood and unambiguous statistic. Standard deviations tend to be relatively stable over time (especially compared to mean returns and other moments of the distribution). Econometricians have developed very powerful tools for accurately forecasting standard deviations.

Critics of the standard deviation point out that it measures the possibility of return both above and below the mean. Most investors would define risk based on small or negative returns (though short sellers have the opposite view). This has generated an alternative risk measure: semivariance of downside risk. Semivariance is defined in analogy to variance, based on deviations from the mean, but using only returns below the mean. If the returns are symmetric–that is, the return is equally likely to be x percent above of x percent below the mean–then the semivariance is just exactly one-half the variance. Authors differ in defining downside risk as the square root of the semivariance, in analogy to the relation between standard deviation and variance. Investors with different preferences may choose different definitions of downside risk.

Downside risk clearly answers the critics of standard deviation by focusing entirely on the undesirable returns. However, there are several problems with downside risk. First, its definition is not as unambiguous as standard deviation or variance, nor are its statistical properties as well known, so it isn't an ideal choice for a universal risk definition. Second, it is computationally challenging for large portfolio construction problems. Third, to the extent that investment returns are reasonably symmetric, most definitions of downside risk are simply proportional to standard deviation or variance and so contain no additional information. To the extent that investment returns may not be symmetric, there are problems forecasting downside risk. Return asymmetries are not stable over time and so are very difficult to forecast. Realized downside risk may not be a good forecast of future downside risk. Moreover, we estimate downside risk with only half of the data, losing statistical accuracy.

Shortfall probability is another risk definition and perhaps one closely related to intuition of what risk is. The shortfall probabilities are the probability that the return will lie below some target amount. Shortfall probability has the advantage of closely corresponding to an intuitive definition of risk. However, it faces the same problems as downside risk: ambiguity, poor statistical understanding, difficulty of forecasting, and dependence on individual investor preferences for shortfall targets. Forecasting is a particularly thorny problem, and it's accentuated for lower shortfall targets. At the extreme, probability forecasts for very large shortfalls are influenced by perhaps only one or two observations.

Value at risk is similar to shortfall probability. Where shortfall probability takes a target return and calculates the probability of returns falling below that, value at risk takes a target probability (e.g., 1 percent or 5 percent lowest returns) and converts that probability to an associated return. Value at risk is closely related to shortfall probability and shares the same advantages and disadvantages. Where does the normal distribution fit into this discussion or risk statistics? The normal distribution is a standard assumption in academic investment research and is a standard distribution throughout statistics.

It is completely defined by its mean and standard deviation. Much research has shown that investments returns do not exactly follow normal distributions but instead have wider distributions– that is, the probability of extreme events is larger for real investments than a normal distribution would imply. Thus, we either need to apply other rules such as the Chebyshev's rule or seek a broader definition and measurement of risk.

I and a colleague (Dr. Smith) have developed such a measure based on tail thickness and have discussed its advantages over the measurements presented above.

The new measure using the parameters k and σ can compute measures of tail thickness, for a given value of the threshold u, that incorporate the characteristics of tail behavior.

One very appealing measure is the conditional mean exceedance (or cme) function,

$$M(u) \equiv E(Z-u \mid Z>u) = E(Y \mid Y>0),$$

which differentiates among different types of upper tail behavior. The cme is the average amount by which the random variable Z exceeds the threshold u given that it is larger than u. Formally, for a GPD,

$$E(Y \mid Y>0) = (\sigma - ku) / (1-k),$$

where, as before, σ denotes the scale parameter for the GPD of the excess Y ($\equiv Z-u$). This simple formal result provides a convenient and meaningful measure of tail thickness for a given value assigned to u. As one can readily observe, the right side (RHS) of this equation is "large" for a "long-tailed" distribution (tail shape parameter $k<0$) and "small" for a "short-tailed" distribution ($k>0$).

Of particular interest is the result that for a "medium-tailed" or exponential distribution (k = 0), the RHS reduces to σ, here denoting the standard deviation of an exponential distribution. The above five risk definitions all attempt to capture the risk inherent in the "true" return distribution. An alternative approach could assume that returns are normally distributed. Then the mean and standard deviation immediately fix the other statistics: downside risk, semivariance, shortfall probability, and value at risk.

Portfolio Risk

Suppose that risk is measured by the standard deviation of the returns. When there are only *two* shares in a portfolio, the variance of the expected portfolio return is given by the formula

$$x_1^2\sigma_1^2 + x_2^2\sigma_2^2 + 2x_1x_2\sigma_1\sigma_2\rho_{12} \tag{1}.$$

Here, x_1 and x_2 are the proportions of the portfolio in shares 1 and 2, σ_1 and σ_2 are the standard deviations of the rate of return for each share, and ρ_{12} is the coefficient of correlation between the returns for share 1 and those for share 2.

This is because the returns of each share, r_1 and r_2 as well as portfolio return, r, are all random variable with the following relationship:

$$r = r_1x_1 + r_2x_2$$

$$Var(r) = Var(r_1x_1 + r_2x_2) = Var(r_1x_1) + Var(r_2x_2) + 2\operatorname{cov}(r_1x_1, r_2x_2)$$

$$= x_1^2Var(r_1) + x_2^2Var(r_2) + 2x_1x_2\operatorname{cov}(r_1, r_2).$$

This formula can be generalized for a portfolio of n shares in the form

$$\sum_{i=1}^{n}\sum_{j=1}^{n}x_ix_j\sigma_i\sigma_j\rho_{ij} \tag{2},$$

where $\rho_{ii} = 1$. As more shares are added to the portfolio, the covariances become more important since the number of terms in expression (2) rises with the square of n (in fact, n^2-n), although the terms not involving ρ_{ij} only rise linearly (in fact n).

Thus, the variability of a highly diversified portfolio reflects mainly the covariances. If the covariance (the term used to describe $\sigma_i \sigma_j \rho_{ij}$) average were zero, it would be possible to eliminate all the risks by holding sufficient shares. Unfortunately, shares move together, not independently. Thus, most of the shares that the investor can actually buy are tied together in a web of positive covariances, which set the limits to diversification. It is the average covariance that constitutes the bedrock of risk remaining after diversification has done its work. Note the risk in minimum when $\rho_{12=-1}$ when there are two shares in a portfolio. When there are three, then negative correlation between share 1 and 2 and also between 1 and 3 implies positive correlation between 2 and 3. If a portfolio consists of n different securities, the proportionate contribution of the jth security to the overall risk of the portfolio is given by the expression

$$x_j \sigma_{jm} / \sigma_m^2 \tag{3},$$

where σ_{jm} is the correlation of the jth security with the market portfolio, σ_m^2 is the variance of the market, and x_j is the proportion by value of the portfolio in share j.

While diversification makes sense for individual investors, it is not automatically true that it is the best course of action for a commercial firm to follow.

Portfolio Diversification and Its Limitations

A first step in examining the diversification effect in portfolios is to consider the characteristics of hypothetical portfolios formed with varying numbers of shares.

For this purpose, *three* simplifying assumptions are made:

(1) All holdings in the portfolio are assumed to be of equal (monetary) size.
(2) All holdings are assumed to be equally risky in the sense described in the previous section.
(3) The risks of each pair of holdings in the portfolio are assumed to be mutually independent.

An examination of price changes in recent years shows that, on average, about *30 percent* of the price movement of a share has been contingent on what has been happening to the market as a whole. Of course, any two shares selected from the same industry group of shares would have had considerably more in common than just this 30 percent. However, since the object of the hypothetical portfolio is to measure the maximum effect that diversification can reasonably be expected to have, it will be assumed that such duplication is never necessary in a portfolio and that the shares in the portfolio have only the market influence in common.

The maximum theoretical benefits from diversification are secured with a portfolio composed of an infinitely large number of holdings.

If the only relationship between any two holdings in a portfolio lies in the fact that 30 percent of each share's prospects is contingent on the behavior of the market, then calculations on the assumptions given above show that no amount of diversification can reduce the risk, or volatility of possible returns, below *74 percent* of that of a one-share portfolio.

Not only is the potential total benefit from diversification limited but also a large part of this potential can be realized with a portfolio of relatively few shares.

A portfolio of *ten* shares provides *88.5 percent* of the possible advantages of infinite diversification; one of *twenty* shares provides *94.2 percent* of these advantages, etc.

Modeling the Stock Market

The stock market cannot survive in a deterministic world.

In recent years, picking stocks in an almost random fashion has become popular among some investors on the grounds that it works as well as expert advice. People familiar with probability theory rely on the finding that what is known as random walk, Brownian motion, and models of this type provide a reasonable description of stock market fluctuations. Although there is general agreement that such models are reasonable, why they work is not entirely understood. Following is an analysis of the general characteristics of these models in an attempt to compare their behavior to the behavior of typical investors. Both deterministic and probabilistic analysis will be discussed, along with some comparison.

Paul Cootner was one of the first economists who long ago recognized the importance of what was then largely statistical analyses of stock price movements and began to provide a theoretical foundation for the apparently random behavior of stock price changes. In his first paper on stock market prices in the '60s, he provided an operational definition of market efficiency and its relation to random walk. Cootner later discussed this in more detail in a collection of essays.

Why Probabilistic Analysis?

Consider a single stock and the people who are interested in it. To study the stock's price changes, we'll pick a starting point and look at the situation as a system. A priori, the choice of initial conditions presents a considerable problem. Imagine just one popular stock and examine the possibility of representing the behavior of investors through some deterministic theory.

1. We would need to know the initial investment and initial direction (investment decisions, plan, motive, and so on) for every investor. This makes the initial data enormous.
2. To perform the calculation, we would need to solve a large system of equations featuring initial conditions. The result would consist of many numbers representing the state of the system at a later time.
3. Finally, we would need to calculate the readily measurable macroscopic performance variables as sums over these output values.

Given the present knowledge and technology, we can see that attempts to implement such a deterministic program must run into several insurmountable obstacles. First, there is the practical impossibility of knowing simultaneously the investment and the direction (plan) taken by individual investors. Next, even if we had this information, such calculations would require a large memory and would be so lengthy as to be impracticable. Finally, there is a methodological objection that one can formulate as follows: since the aim is to calculate macroscopically measurable quantities from microscopic data, might one not do this more easily in some other fashion than a deterministic analysis?

The story is similar to what happened in physics when Maxwell in 1859 could not but realize that in practice, deterministic calculation related to molecules of even one mole of gas is impossible. Here, the initial investment and direction taken by an individual are similar to the initial position and velocity of a molecule. It was in this context that Maxwell made the conceptual leap of introducing into the theory the notion of unpredictability. His breathtaking intuition was developed further by Boltzmann and then confirmed half a century later by the work of Albert Einstein in 1905 and of Jean Perrin in 1908 on Brownian motion. It is important to note that Maxwell does not reject determinism as such, but he mitigates our ignorance of the initial conditions by introducing assumptions of probabilistic nature. His approach can be placed within a framework of the so-called probabilities through ignorance. The introduction of this approach in 1872 led to the birth of statistical physics.

In a similar fashion, when modeling the stock market, one can fruitfully exploit probabilities that represent our ignorance. Basically, we admit the deterministic nature of actions taken by investors but try to find a way to treat them probabilistically. In other words, to construct the economic theory governing a deterministic system of large number of investors, it is possible to exploit probabilities through ignorance with complete success. In sum, based on this discussion, the possible models may be described in the following way:

1. Probabilistic models use randomness as a basic concept so that market is governed, at least partially, by chance and associated laws of probability.

2. Chaos offers the fascinating possibility of describing randomness as the result of a known deterministic market. Randomness may lie on the choice of initial conditions and the subsequent mechanism of the market change. Note that the term "chaos" is usually reserved for dynamical systems whose state can be described with differential equations in continuous time or difference equations in discrete time. This means that it is possible to represent the market using these models.

References

Noubary, Reza. 2001. "The Stock Market: Deterministic or Random?" In *Applied Statistical Science*, edited by V. M. Ahsanullah, J. Kennyon, and S. K. Sarkar, 261–266. Nova Science.

Chaos: A Misunderstood Concept

Unlike popular understanding, there is an order in chaos.

Chaos is a subject that usually fascinates both mathematicians and nonmathematicians. However, what they understand of it may be very different. Often, people think of chaos has no order or is all noise and no signal. This is unfortunate since it actually refers to a high degree of order. As pointed out by Joseph Malkevitch in his article "Mathematics and Climate," we are used to noise of various kinds: static on the radio, conversations at other tables in a crowded restaurant, or the hums one hears from appliances around the house. In all of these situations, there is a "randomness" or chance element in what we experience. The hum is not a single tone but is constantly changing. The static on the radio comes and goes in frequency and intensity. Thus, everybody was surprised when it became widely known that seemingly "noisy" situations arose from systems that involved nothing random but were completely deterministic. This phenomenon is now known as "chaos" or "chaotic behavior." The mathematical roots of insight into deterministic chaos go back to the great French mathematician Henri Poincaré. However, in more modern times, one early pioneer in noticing chaos was the meteorologist Edward Lorenz, who lived from 1917 to 2008. Lorenz started his academic career by studying mathematics as an undergraduate at Dartmouth and earned a master's degree in mathematics from Harvard. Later, he received a doctorate from MIT in meteorology with a thesis titled "A Method of Applying the Hydrodynamic and Thermodynamic Equations to Atmospheric Models," where he put his mathematical training to work. Lorenz came across the phenomenon now known as chaos in conjunction with the study of the behavior of the solution of differential equations.

Chaos and Stocks

In physics, using Maxwell's idea, one can show how chance could generate determinism. Here, we look at a situation where determinism can generate chance and points out the way the resulting theory may be considered appropriate for stock market analysis. Chaos is a particularly unfortunate name because, unlike what we understand from the word, it actually refers to a higher degree of order. To appreciate its importance, we can refer to the fact that the heart has to be largely regular or we die. But the brain has to be largely irregular–if not, we have epilepsy. This shows that irregularity, chaos, leads to complex systems. It is not all disorder. As pointed out earlier, chaos theory was pioneered near the end of the last century, but only the advent of fast computers in the early 1960s has made its development possible. Today, it is a very active field of research and has induced a far-reaching revolution in our concepts. To classify it, we make a semantic distinction between random and chaotic processes.

A dust particle suspended in water moves around randomly, executing what is called Brownian motion. This stems from molecular agitation through the impacts of water molecules on the dust particle. Every molecule, much like an investor, is a direct or indirect cause of the motion, and we can say that the Brownian motion of the dust particle is governed by many variables. In such cases, one speaks of a *random process;* to treat it mathematically, we use the calculus of probabilities.

A compass needle acted on simultaneously by a fixed and by a rotating field constitutes a very simple physical system pending on only three variables. However, one can choose experimental conditions under which the motion of the magnetized needle is so unsystematic that prediction seems totally impossible. In such very simple cases whose evolution is nevertheless unpredictable, one speaks of *chaos* and of *chaotic processes;* these are the terms one uses whenever the variables characterizing the system are few.

Thus, when the system such as the one we based on behavior of individual investors is studied, one looks at a very large number of variables, and hence, random process is more appropriate. However, when only a few variables such as interest rate and currency rate can characterize the system, the chaos may be used.

The relevance of chaos to market behavior may also be explained by the following example. A pendulum swings to and from with a regular back-and-forth motion, but if it is struck by the ball of a second pendulum before reaching its zenith, both pendulums may begin swinging in wildly erratic patters. In the financial market, a trend is enhanced or undermined by surprises in governmental announcements or economic actions by one or more influential nations.

References

http://ams.org/publicoutreach/feature-column/fcarc-climate.

Compound Interest, a Powerful Force

An account starts with $1.00 and pays 100 percent interest per year. If the interest is credited once, at the end of the year, the value of the account at year-end doubles to $2.00. What happens if the interest is computed and credited more frequently over the course of the year? If the interest is credited twice in the year, the interest rate for each six months will be 50 percent, so after the first six months of the year, the initial $1.00 is multiplied by 1.5 to yield $1.50. Reinvesting this by the end of the year, it becomes $1.50 x $1.50, yielding $1.00×1.5^2 = $2.25 at the end of the year. Compounding quarterly yields $1.00×1.25^4 = $2.4414 . . . and compounding monthly yields $1.00 × (1+1/12)^{12}

= \$2.613035. Compounding weekly (n = 52) yields \$2.692597 . . ., while compounding daily (n = 365) yields \$2.714567 . . ., just two cents more. The limit is the number that came to be known as e; with continuous compounding, the account value will reach \$2.7182818. Bernoulli, in 1683, was the first person to notice this.

An article titled "Does the number e have any real physical meaning, or is it just a mathematical convenience?" discusses whether the number e has real physical meaning or if it is just a mathematical convenience. According to the article, the answer is, yes, it has a physical meaning. It occurs naturally in any situation where a quantity increases at a rate proportional to its value, such as a bank account producing interest or a population increasing as its members reproduce. Obviously, the quantity will increase more if the increase is based on the total current quantity (including previous increases) than if it is only based on the original quantity (with previous increases not counted). How much more? The number e answers this question. To put it another way, the number e is related to the how much more money you will earn under compound interest than you would under simple interest.

The more mathematics and science you encounter, the more you run into the number e. Since its discovery, it has shown up in a variety of useful applications including (but definitely not limited to) solving for voltages, charge buildups and currents in dynamic electrical circuits, spring/damping problems, growth and decay problems, Newton's laws of cooling and heating, plane waves, and compound interest; e also helps determine the best applicant for a job and waiting time to the new record and many more.

References

Blank, Brian. "A Joint Review of *A History of Pi*, by Petr Beckmann, St. Martins's Press, 1976, Barnes and Noble Books, 1989; *The Joy of Pi*, by David Blatner, Walker & Co., 1997; *The Nothing That Is*, by Robert Kaplan, Oxford University Press, 1999; *e: The Story of a Number*, by Eli Maor, Princeton

University Press, 1998; *An Imaginary Tale*, by Paul Nahin, Princeton University Press, 1998; *Zero: The Biography of a Dangerous Idea*, by Charles Seife, Viking Press, 2000." http:// delivery.acm.org/10.1145/1170000/1165560/p19-blank. pdf?ip=148.137.59.58&id=1165560&acc=ACTIVE%20 SERVICE&key=A792924B58C015C1%2EC80AEB80C46F BF65%2E4D4702B0C3E38B35%2E4D4702B0C3E38B35&__ acm__=1539619085_72d5be5631b46e1665678f52320ac548.

"Does the Number e Have Any Real Physical Meaning, or Is It Just a Mathematical Convenience?" http://www.math.utoronto. ca/mathnet/answers/ereal.html.

"e (Mathematical Constant)." https://en.wikipedia.org/wiki/E_ (mathematical_constant)#cite_note-OConnor-5.

Ellinor, Andrew, Patrick Corn and Geoff Pilling. "The Discovery of the Number e." https://brilliant.org/wiki/ the-discovery-of-the-number-e/.

CHAPTER 6

Risk and Education

Glad that some monkeys took risks to become men.

ALTHOUGH I HAVE been teaching introductory probability and statistics for more than forty-five years, I still find the task challenging. To help this, I try to take basic concepts and tie them to everyday problems so that the subject matter becomes meaningful to students. Recently, I designed and implemented an experiment using an idea for motivating students in an introductory probability and statistics course. The selected theme was risk, and the process started with a first-day-of-class questionnaire (presented below), which sampled attitudes of students toward risk and involved them in analysis of events and decisions from their daily lives. Questionnaire responses served as a context for developing the technical concepts of probability and statistics and for increasing students' motivation and involvement in the course. I selected this theme since my earlier research indicated an increased student involvement and interest in statistics. My plan is to add data related to pandemics that all of us including students had to deal with.

Significance

Reluctant students ask why they should study probability and statistics. Our answers each time have been a variation on the theme, "to learn how to make sensible decisions in the face of uncertainty." Many textbooks try to motivate students by introduction of varied applications. This technique addresses students' apparent desire to see the relevance of their studies to the outside world and their skepticism about whether statistics has value. In recent years, several good textbooks have been published that are based on the belief that applications–to students' everyday experiences or to their chosen career fields–are the key to motivation. For our introductory students, this approach often fails to motivate. It may work with students who, beforehand, are deeply committed to a particular academic or career field, for they can select and investigate research questions that interest them. However, even for these committed students, applied examples and exercises may fail to motivate because they are introduced after the presentation of complex technical concepts; by then, some students have already turned against statistics. Moreover, these examples are not usually of immediate concern to the majority of students since they lie too far away into a student's future instead of relating to their present experiences. For example, sophomores who plan to be teachers are not highly motivated by statistical experiments that compare different teaching styles; this lack of motivation occurs even among those who are highly committed to a career in education. As sophomores, they are simply too far away from their future teaching experience to be excited by examples of this sort.

Several investigations have discussed the importance of using real data to improve the teaching of applied statistics. I found that most of the students who enroll in our introductory statistics courses are not fascinated by real data and its careful analysis. Perhaps this occurs because the data typically comes from discipline such as economics, education, psychology, or yet another field in which immature students do not yet have a deep interest.

Another frequently made suggestion is to involve students in data collection and its analysis using a computer. Most of our students take only one statistics course, and one semester does not provide enough time for proper implementation of this idea. I found that computer instruction steals time from statistical topics, and students lack enthusiasm for data collection because of problems related to sampling. Additionally, students find that statistics applied to someone else's problems seems dead and dull rather than useful and interesting.

However, the prospects for motivating students are not entirely bleak. Despite many voids in student interests, despite many ideas that fail to motivate them, they do have common interests that we can build on as we try to teach them statistics. People of all ages perform experiments, evaluate uncertainties, and analyze information as an integral part of everyday life. This occurs whether or not they have studied statistics. As their teachers, we could relate statistics to their student experiences and help them to use the subject to become more effective decision makers. Based on my experience, linking their studies to a familiar subject with which they have immediate concern almost always works better.

Goals and Objectives

Using risk is an approach different from each of the motivational strategies that may have been found in popular textbooks. It suggests that motivation of students to study probability and statistics can be accomplished by linking its study to an idea that they have dealt with all their lives–namely, risk. With this focus, the instructor can help students to build their study of statistics on a foundation–an understanding of risk–already possessed by all students. Throughout the course, students and instructor can work together to enlarge and clarify student understanding of this concept using the basic ideas of probability and statistics. Moreover, this approach (which I experimented using college students, primarily at the sophomore level) is adaptable at all levels at which statistics is taught, from junior high to professional schools.

This idea is one that has been germinating in me for some time, and during the last few years, risk has finally grown into a full-grown theme. Since my study produced results that, despite its weakness and inconsistencies, led to a publication in a reputable journal, I am very hopeful that a further research on statistics education as well as risk theory will produce results useful to professionals in general and to the instructors teaching statistics in particular.

Methodology

The following is a description of the methodology that I implemented. The risk-centered approach will begin with a questionnaire, like the one shown below. On the first day of class, the professor will give each student the assignment of completing this questionnaire (spread over several pages with space for thoughtful answers) for the next class period. To ensure that students will participate in this activity, he or she will promise to count it as a quiz. When handing out the questionnaire, he will ask students to read it carefully and make sure that they understand each question; he will speak with students individually and offer clarifications.

Throughout the semester, the professor will refer to particular groups of answers and examples provided by students on their questionnaires. In my pilot study, this led to a high level of interaction among students and between students and the professor; sometimes the discussions were heated as students defended differing points of view.

At the end of the semester, the professor will utilize methods developed in this project to compare each class with which the questionnaire is used with previous non-questionnaire classes. The results of study showed that not only had student involvement and motivation been high, but also a significantly higher percentage of students in questionnaire-using classes elected to take a second statistics course and some even decided to minor in statistics.

Statistics: Opening-Day Questionnaire

1. Risk is an everyday problem. What do you know about it? Explain.
2. Which of the following is riskier? Explain.
   (a) car   train   plane
   (b) smoking   x-ray   nuclear power plant

3. How do you deal with risk? (Try to avoid? Buy insurance? Enjoy it? . . .)
4. Do you think that the following statement is correct?
   Not taking a risk is taking a risk of another kind.
   Explain.

5. Have you ever been involved in any risky activities? Describe one case.
6. Have you ever made a decision that involved risk? Describe one case.
7. For your decision in question 6, how much risk was involved?
8. How did you arrive at your risk-estimate in question 6?
   (a) If your answer to question 7 is a word like "high" or "low" rather than a number, can you express it numerically using percentage? Explain.

   (b) Order the items in question 2(a) from the least risky to the riskiest/most risky. If the risk of the least risky item is r, then what should be the risks of the other items (e.g., 2r? 3r? . . .)?

9. What kinds of background information does a person need to make reasonable numerical estimates of the risks that you mentioned in questions 6 and 7?

10. What branches of science are involved in the process of estimating risk? What role do you see for "probability and statistics" in this task?

11. Look back and summarize your answers: overall, how do you measure risk?

12. Which of the following definitions of risk do you prefer? Explain.

(a) Risk is the chance of loss.

(b) Risk is the possibility of loss.

(c) Risk is uncertainty.

(d) Risk is the dispersion of actual from expected results.

(e) Risk is the probability of any outcome different from the one expected.

(f) Risk is the probability of an event times the probable cost (or loss) if the event occurs.

13. What do all of the definitions in question 12 have in common? What role does "probability and statistics" play in each? Explain.

14. Do you agree with each of the following statements? Explain why or why not.

(a) Risk may be objective or subjective.

(b) Quantifying risk means determining all the possible outcomes of a risky activity and determining the relative likelihood of each outcome.

15. Suppose that you are thinking about flipping a fair coin. To quantify this, which of the following approaches do you prefer?

   (a) Repeat the flip a large number of times until you have established the result that half of the time it comes up tails and half of the time heads.

   (b) Assign 50-50 likelihoods since there are only two possible outcomes and you have no reason to say that they are not equally likely.

16. Have any of the later questions changed your answer to early questions?

17. Can you see ways that learning probability and statistics will enable you to come with more accurate answers to these questions? Explain.

Description of the Method

The following is a description of the method used. The risk-centered approach began with a questionnaire presented earlier. On the first day of class, the professor gave each student the assignment of completing this questionnaire (spread over four pages with space for thoughtful answers) for the next class period. To ensure that students would carefully participate in this activity, he promised to count it as a quiz. When handing out the questionnaire, he asked students to read it carefully and to make sure that they understood each question; he spoke with students individually and offered clarifications.

We make no special claims for the questionnaire itself. It was constructed quickly, and it turned out that a couple of the questions were misunderstood by some students. In addition, several questions elicited almost identical answers. Despite these factors, responses from about 150 students gave us a rather clear picture of the degree of their naiveté about probability and statistics and, more importantly, provided us with information from students' day-to-day lives to which we can help them attach the concepts from probability and statistics that we seek to teach.

Throughout the semester, the professor referred to particular groups of answers and examples provided by students on their questionnaires. This led to a high level of interaction among students and between students and the professor; sometimes the discussions were heated as students defended differing points of view. These differences provided the perfect opportunity for the professor to point out that in probability and statistics, there are no single unified agreed-upon definitions of either "probability" or "risk."

At semester's end, as the professor compared each class with which the questionnaire was used with previous nonquestionnaire classes, the results were remarkable. Not only had student involvement and motivation been high, but also a significantly higher percentage of students in questionnaire-using classes elected to take a second statistics course. Needless to say, we are pleased with these results and would like to do more research and investigation.

The instructor will be responsible for revising the questionnaire and making it more consistent and for collecting the data and improving the ways student responses may be used for introducing the key statistical concepts. She will evaluate the findings and prepare the final report.

Can We Be Sensible about Risks?

This section examines some current attitudes toward risk with the goal of educating ourselves and others. Its objective is to convey some of the complexity that is involved in assessing risk and to provide background that will help the decision-making in future risk assessment.

In day-to-day conversation, we make statements like "Mary is a risk-taker" or "John is not a risk-taker" to describe the attitudes of people we know toward actions that may lead to loss or damage. We take a risk when we cannot predict or control fully the consequences of an action.

Acceptance of risk by individuals is often very different from the acceptance of the same type of risk by groups. For risks that involve ourselves alone, we may feel comfortable with subjective decision-making, relying on hunches or gut feelings; however, groups often must justify their decisions to others and thus may need to make decisions on objective bases. Individually, we take risks when we decide whether to get married or to smoke or when to cross the road. Communities also take risks such as when they build roads or protection against flooding or when they fluoridate the water supply. Often, these decisions are based on some quantitative analysis. They may decide not to protect against flooding if the costs are too great; they may decide to fluoridate the water supply because they believe that the good effects outweigh the bad ones.

Risk, as discussed earlier, has two components: the uncertainty of the occurrence of an undesirable event, often expressed as a probability; and the variable magnitude of its consequence, often measured in dollars. Both components are difficult to estimate. In what follows, we examine some current attitudes toward risk with the objective of providing a background that will help the reader in future risk assessment.

Television, reflecting great public interest and concern, provides extensive coverage of natural disasters with strong visual impact. Yet television is often criticized for the way it presents risks. Though progress has been made in recent years, still assessing the role of the broadcast media in elaborating and reinforcing dominant public perceptions of risks during disasters is needed.

Experts versus Public

In today's world, there is much disagreement about which risks are of greatest concern. For example, nuclear power is considered a major risk by members of the public; experts, on the other hand, consider it riskier to travel by train. This disagreement is clearly seen in table 1, which gives an ordering of the risks that people see associated with twenty different activities.

First note that, here, risk means "mathematical expectation." For example, if we face an outcome that has a numerical value–such as winning $400 in a raffle–and if we multiply the probability of the outcome by the value of the outcome, that value is our expectation. If we have one chance in one thousand of winning $400 in a raffle, then our expectation is

$$(0.001) \times \$400 = \$0.40.$$

Thus, the expectation of anyone holding a lottery ticket is 40 cents. Another way of looking at this value is as the "average" or mean amount won by all of 1,000 ticket holders, one of whom wins $400 and 999 of whom win nothing.

Consider the following table. We observe that even groups of educated, relatively well-informed individuals (college students and members of the League of Women Voters) differ from the "experts."

Ordering of Perceived Risk for Twenty Activities and Technologies

	College Students	League of Women Voters	Experts
Nuclear power	1	1	20
Handguns	2	3	4
Smoking	3	4	2
Pesticides	4	9	8
Motor vehicles	5	2	1
Motorcycles	6	5	6
Alcoholic beverages	7	6	3
Police work	8	8	17
Contraceptives	9	17	11
Firefighting	10	11	18
Surgery	11	10	5
Food preservatives	12	20	14
Large construction	13	12	13
General (private) aviation	14	7	12
Commercial aviation	15	14	16
X-rays	16	18	7
Electric power (non-nuclear)	17	15	9
Railroads	18	19	19
Bicycles	19	13	15
Swimming	20	16	10

"Experts" and the public often disagree in risk assessment. The table from *The Business of Risk* (1983), by Peter G. Moore, illustrates how groups of people rank the risks associated with certain activities and technologies. Even groups of educated, well-informed individuals differ from those with expertise in risk assessment.

Lack of agreement, shown in the table, between the "experts" and the others is supported by our calculation of the correlations for the ratings for the different groups; these are displayed in table below.

	League of Women Voters	College Students
College students	84%	
Experts	35%	34%

Rank Correlations for Ratings

Who are the "experts" here? Primarily, they are persons with expertise in risk assessment–that is, people who treat risk systematically as an expectation and who have had experience in estimating and working with probabilities. Although they gather extensive information to use in their assessment, the experts are not, as a rule, scientists or practitioners in a field of knowledge– such as nuclear power or environmental science in which major risks have been identified. Their expertise lies in the process of using available information to assess risks.

Are the experts right? On the one hand, there is plenty of evidence on which to base an assessment of the risks of train travel, but there has never been a total nuclear disaster. Just what are the chances that it will occur? Experts might use information like that presented in table below, which is based on data collected in the United States and shows the annual death rates for people killed in various ways within twenty-five miles of a nuclear reactor (Moore 1983).

Annual death rates in the neighborhood of a power plant

Type of accident	Death rate
Car accident	1 in 4,000
Accidental fall	1 in 10,000
Fire	1 in 27,000
Reactor accident	1 in 750,000

Estimation of Probabilities

How does an informed member of the general public estimate probabilities of various risks? According to some publications, Americans are innumerate and, in particular, woefully poor at using numbers meaningfully to make estimates about the objects and events in the world around them. However, some suggest a different view—namely, that errors in probability estimates are based on biased coverage of events by the media.

Combs and Slavic examined the coverage by newspapers in New Bedford, Massachusetts, and Eugene, Oregon, and also surveyed readers of these papers. The people whom they surveyed had attitudes toward causes of death that were directly related to the amount of coverage of various types of events in their local newspapers. In fact, these patterns emerged:

Although diseases take about sixteen times as many lives as accidents, the newspapers contained more than three times as many articles about accidents as about diseases, noting about seven times as many accidental deaths.

Although diseases claim almost one hundred times as many lives as homicides do, there were about three times as many articles about homicides than there were about disease deaths and each twice longer.

The people who read these newspapers assessed the risk of death by homicide as much greater than the risk of death by accident, which, in turn, was thought to be much greater than the risk of death by disease. Although their assessment had an incorrect result, nevertheless, these readers made interpretations consistent with the numbers of the deaths about which they read. In other words, the readers' estimation processes were correct, but the results were incorrect because they relied on incomplete information.

In a *Scientific American* article that contains numerous thought-provoking examples, Daniel Kahneman and Amos Tversky (1982) explore at length the role that the framing of a choice plays in the resulting decision. Compare, for example, the following pairs of choices:

CHOICE 1

On the table in front of you is $200; you will get it and possibly something more. To determine how much more, you must choose between:

A: avoid a gamble and collect $50 more for sure;
B: gamble with a 25 percent chance of winning $200 more and a 75 percent chance of winning nothing more.

Do you prefer A or B?

CHOICE 2

On the table in front of you is $400; you will get it and possibly something more. To determine how much more, you must choose between:

C: avoid a gamble and lose $150 for sure;
D: gamble with a 75 percent chance of losing $200 and a 25 percent chance of losing nothing.

Do you prefer C or D?

According to Kahneman and Tversky, most people prefer A to B and D to C, even though the expected gains are identical for A and C and also for B and D.

Choice B has the expectation

$(0.25) \times \$200 = \$50$, to be added to the initial $200 yielding $250,

and choice D has the expectation

(0.75) X \$200 = \$150 to be subtracted from the initial \$400 yielding \$250. Thus, both choices lead to the same expectation of \$250.

Consider also the following example in which we see several different ways of looking at the risk of not winning in a lottery:

The winning ticket in a lottery was 865304. Three individuals compare the ticket they hold to the winning number.

> John holds 361204; Mary holds 965304; and Peter
> holds 865305.

If one million different tickets have been sold, then every ticket holder is very likely not to win, and each just as likely as the others. The chance of winning is only one in a million. Why, then, is Peter very upset about losing, even more so than Mary, and John is only mildly disappointed?

We may suppose that the ticket holders' disappointments are related not only to winning or losing, since all have lost, but also to the chance of matching the winning ticket. Only fifty-four of the one million tickets match the winner as closely as do Peter's and Mary's tickets, differing only in a single digit. John's ticket differs from the winner in half of its six digits, and, thus, he has not "come close" to winning. A question still remains: why should Peter be more disappointed than Mary? Is he thinking not of a single winning ticket but of a list of digits drawn in order? Reading the digits from left to right, one sees at once that Mary is not a winner but does not know that the same is true for Peter until the end.

References

Tversky, A., and D. Kahneman. 1982. "The Psychology of Preferences." *Scientific American*, 246(1): 160–173.

Paradoxes Involving Probabilities

To reason correctly when we use probabilities is difficult and requires thinking that seems contrary to intuition. For example, consider the following hypothetical situation concerning a diagnostic test for cancer. Suppose that in the over-thirty-year-old population in the United States, 0.5 percent, one out of two hundred, people actually have cancer. Suppose further there is a test for cancer that is 98 percent accurate: if the test is performed on persons who have cancer, the test results will be positive 98 percent of the time, and if the test is performed on persons who don't have it, the test results will be negative 98 percent of the time. Table below shows the expected results if this test is performed on ten thousand people (of whom about fifty actually will have cancer).

Hypothetical results of 98 percent accurate cancer detection test

	Number of positive test results	Number of negative test results
People free of cancer (9950)	199	9751
People having cancer (50)	49	1
TOTAL	248	9752

Suppose you have had the test and the results come back positive. Do you have cancer? Probably you would reason that since the test is 98 percent accurate, it is almost certain that you do.

However, a positive test result is wrong for most people and you probably would not have cancer. Look back at table above. What portion of the persons with positive test results actually have cancer? The answer is 49/248, or slightly less than 20 percent. Although the test is 98 percent accurate, since most people are healthy, most of the test's errors are on healthy people, and a positive test is likely to be in error.

Here is a second counterintuitive probability example. This one deals with the unlikely event that two randomly chosen people will share the same birthday. Surprisingly, the likelihood of coincident birthdays becomes high in groups that are much smaller than 365.

A particular professor with an announced policy of occasional surprise quizzes comes in to meet her class of thirty students and asks each student to write down his or her birthday month and day, ignoring the year, on a slip of paper. She collects the slips and then announces that there will be a surprise quiz today if two of the slips name the same birthday. As preparations are being made to tally the dates, do you suspect that the students are already beginning to relax, confident that with only thirty students and 365 days to choose from, there will not be a match?

Although we cannot say for certain what will happen in this particular class, elementary concepts of probability theory and a straightforward calculation can be used to discover that the probability of a repeated birthday in a group of thirty people (with no twins and no February 29 birthdays) exceeds 70 percent. In fact, the probability of a repeated birthday is over 50 percent for a group as small as twenty-three people.

There is perhaps no more appropriate way to end a consideration of the pitfalls inherent in working with probabilities than with a quote from *The Bending of the Bough,* by George Moore: "There is always a right way and a wrong way, and the wrong way always seems the more reasonable."

Eliminating Risks

Almost any risk can be eliminated if we are willing to pay the price. If you pay $300 per year to insure your $80,000 home against loss by fire, and if there is only one chance in two thousand that this loss will occur, then you might evaluate this situation using the following procedure:

Imagine two thousand people like yourself paying $300 per year to insure their $80,000 homes against loss by fire. They pay a total of $600,000 in premiums, but only one collects $80,000. The insurance company thus is collecting $520,000 ($260 per policyholder) more than it is paying out for this loss. (Of course, the costs of operating an insurance company and small claims will reduce this profit somewhat.) Looking at it from the insured's point of view, most people lose money each year on fire insurance. Is purchase of fire insurance, then, an unwise decision?

The decision to purchase insurance can be evaluated using mathematical expectation. Using positive numbers to denote gains and negative numbers to denote losses, we multiply each consequence by its probability and add them:

$$(0.0005) \times (\$80,0000 - \$300) + (0.9995) \times (-\$300) = -\$260.$$

Thus, the mathematical expectation from purchasing this a fire insurance policy is a loss of $260. This figure agrees with our observation above that the insurance company is collecting $260 more per policyholder than it is paying out. Rather than being foolish when they purchase insurance, people are making a trade, accepting a loss of money in exchange for the "peace of mind" that comes from protection against the very unlikely but very devastating loss of a home.

We may also be paying for an intangible item, the pleasure of dreaming, when we buy lottery tickets. Even those of us who know how slim our chances of winning are and who know that our mathematical expectation from the purchase of a fifty-cent or one-dollar ticket is only a few cents still occasionally buy tickets. Although we know that the purchase of a ticket is almost equivalent to throwing the money away, the attractiveness of the prize makes us dream, and we trade the price of a ticket for the pleasure of holding the dream for a while.

What Needs to Be Done?

In the sections above, we tried to convey some of the complexity that is involved in assessing risk. What then, you may wonder, is an intelligent, sensible person to do to increase his or her understanding of risk and to evaluate it properly?

First, we must accept the discipline of a uniform definition of risk. One might compare risky situations sensibly by considering the probabilities of each of the risks; likewise, one might compare them by considering the consequences of each. However, to compare a variety of risks–including ones with enormous consequences but low probabilities and ones with mild consequences but high probabilities–experts evaluate risk as a product of a probability and a consequence.

We cannot make sensible decisions about risks if we do not have accurate information. Media reporting often spotlights spectacular events and ignores commonplace ones. Readily available references such as the annual *World Almanac and Book of Facts* may be used as a source of accurate statistics about death and accident rates, for example.

The birthday problem and the cancer detection test remind us that correct thinking about probabilities can seem paradoxical, running counter to our intuition. In the eighteenth century, the mathematician James Bernoulli began a campaign advocating basic knowledge of probability for all. Perhaps, eventually, our concern about risk assessment will motivate us to reach Bernoulli's goal. Acceptance of the idea of risk as quantifiable, accompanied by a belief that the risks of disasters can be reduced, and perhaps eliminated, is a modern approach to risk that is gradually gaining acceptance, replacing the belief that disasters are caused by the wrath of the gods, who must be appeased by sacrifices. Current scientific and medical practices even attempt to alter the risks associated with the survival selection processes of nature.

One successful risk-elimination project was conducted by the World Health Organization; their campaign to eradicate smallpox cost about $300 million and is saving an estimated two million lives per year, or twenty million lives in ten years. Whatever cash value is assigned to a human life, it surely is more than $15 (the quotient of $300 million and twenty million). Often, risk elimination is not so easily costed out. For some environmental risks such as global warming, by the time we know what the costs will be—either to eliminate the risk or to live with it—it may be too late to do anything. Choices to eliminate risks are very difficult to make, but as a minimum requirement, the following cost-benefit relationship should hold:

Cost of avoiding risk	<	Probability of risk	x	Cost of consequence

Other reasonable quantitative principles also are commonly used in decision-making; however, it would be incorrect to suppose that all decisions are, or even should be, made on a quantitative basis. There are often personal factors, such as "peace of mind" and other pressures such as satisfying the demands of an electorate, on decision makers.

As we learn more about risk assessment, we also learn that the assessment itself is a risky matter. Because both probabilities and costs may be difficult to estimate, our assessment is likely to be wrong. The "experts" persist, believing—contrary to popular wisdom—that a little knowledge is more valuable than dangerous.

Choices

Some people take big risks—a gambler rolling the dice, a wildcatter drilling for oil, a tightrope walker venturing a tentative first step. But experts argue that all of us take risks all the time that risk is inherent in almost all ordinary activities. As we eat our buttered toast, we ingest cholesterol. As we walk through the autumn woods, we breathe polluted air. We take a risk each time we do something for which we cannot predict or fully control the consequences.

Experts–people who treat risk systematically, people with experience in estimating probabilities–understand that a degree of risk may or may not be easily measured; it may be objective or subjective. They consider it an objective risk to call out "heads" or "tails" as someone flips a quarter, because mathematicians have established standard probabilities for flipped coins. While even experts cannot predict the outcome of any particular flip of the coin, they understand the odds they are facing in making a call.

Experts disagree in their judgment of complex environmental risks, such as whether burned waste will produce damaging amounts of toxic substances. From the same data, one researcher might estimate the probability of damage at 1 percent, another at 3 percent. Both would describe any assessment of that risk as subjective. Each takes the available information and provides a number that describes a degree of belief that an event will happen. Although they gather extensive information to use in their assessments, these experts in risk assessment often are not scientists or specialists in environmental matters. Their expertise lies in their skill in working with available information to assess risks.

Many risks, they note, are not so much personally chosen (diet drinks, oral contraceptives) as socially imposed (nuclear power, hazardous wastes). And some of the social risks–those to public health or safety–have become major political issues, provoking widespread uneasiness about scientific progress. We value the products of modem chemistry yet are preoccupied with chemical contamination. We use electricity from nuclear power plants yet fear the prospects of a nuclear accident. We have trouble telling at what point a hazard (something that may cause harm) could tum into a disaster (the actual harm caused by a hazard). We have trouble judging the relative benefits and dangers of the risks we face.

Risk Management

Each of us develop our personal strategy to deal with risk. Some risks we can transfer or eliminate if we are willing or able to pay the price. The growth of the insurance industry shows the widespread effort by people to band together to manage risk by sharing it. We might pay $300 a year to insure our $90,000 home against loss by fire. If there is statistically one chance in two thousand that our home will bum to the ground, we might evaluate the cost of avoiding a loss by fire using this (somewhat simplistic) procedure:

If two thousand people pay $300 per year to insure their $90,000 homes against financial loss by fire, together they pay a total of $600,000 in premiums. But if only one home bums to the ground, only one of the two thousand collects $90,000. For the insurance company, this is the same as paying $45 (i.e., $90,000 divided by 2,000) to each homeowner from whom they have collected $300. All of the people "lose" their $300 premium, but all are protected against losing more. The insurance company is left with $510,000, out of which comes operating expenses and the costs of smaller fires or other insured losses.

Is the purchase of fire insurance an unwise decision? Probably not. Rather than being foolish when they purchase insurance, people are making a trade-off–accepting the loss of a comparatively small amount of money in exchange for the peace of mind that comes from protection against the unlikely but potentially devastating loss of a home.

When most people consider whether to insure their homes against loss by fire or other catastrophes, they analyze the matter only enough to decide whether they can afford the premium that will let them avoid the risk. A more complex consideration is whether to buy a policy that excludes $250 or $1,000 in deductible costs. Some people will pay for the most nearly complete protection they can afford. Others conclude that with only a little risk, they can save money—perhaps for managing other risks—by carrying a less expensive policy with a higher deductible.

Not all risks can be easily avoided. Sometimes simply choosing a career or opening a savings account carries dangers. There may be no jobs for nurses four years from now. A bank could fail, or low interest rates could mean that the value of money in a savings account does not keep up with inflation. Many other decisions, short or long term, could be risky, and the degree of risk could be difficult to assess. San Francisco might have an earthquake. Should we reconsider our trip? A hole in the ozone layer might threaten our health. Should we stay indoors, wear a sun block, move to New Zealand?

We can avoid some risks and learn to manage some. But which ones are most important to manage? How much will it cost to manage them? What trade-offs must we make in dealing with one to have resources left to take on the others? The risks associated with hazardous waste sites offer a good example of the difficulties we face.

The Popular Media

Some evidence shows public perceptions about risks may stem from the way people evaluate media coverage of the news. One recent research project examined how newspapers in New Bedford, Massachusetts, and Eugene, Oregon, reported deaths. It then surveyed readers of these papers and concluded that readers had attitudes toward the causes of death that were directly related to the coverage of various types of deaths in the newspapers.

Although diseases nationally take about sixteen times as many lives as accidents do, the newspapers in New Bedford and Eugene printed over three times as many articles about accidents as about diseases, and seven times as many articles about accidental deaths as about deaths from diseases. While diseases nationally claim almost one hundred times as many lives as do homicides, the newspapers ran about three times as many articles about homicides as about deaths from diseases. And homicide articles were more than twice as long as articles about deaths from disease and accidents.

The people who read these newspapers assessed the risk of death by homicide as much greater than the risk of death by an accident, which, in turn, they thought much greater than the risk of death by disease. They made judgments that seemed consistent with what they read. They relied on information seeming to suggest that homicides and accidents happened more frequently than did deaths from diseases because the newspapers reported on them more often. They confused numbers of stories with numbers of deaths.

Perhaps such an example can tell us something about media coverage of the risks associated with hazardous waste sites. No doubt, the media has served us well in focusing our attention on a problem long neglected. But perhaps treatment of the subject has resulted in people misjudging the relative benefits and dangers we face.

Choices

Some risks we can reduce or eliminate— others we have to live with.

How, then, can we understand risk and evaluate it properly? First, we need accurate quantitative information. We need to want to be informed, to seek out accurate information, to support efforts to compile useful data. Publications have, for years, offered expert advice on assessing some risks. Consumer Reports, known mostly for estimating the reliability of a car or the quality of sound produced by a loudspeaker, has been turning attention recently to health and environmental concerns. Pamphlets put out by the United States Government Printing Office offer advice on hundreds of subjects, from choosing a mortgage to testing for radon gas.

We can find accurate data in the popular news media as well, though most journalists report on immediate, spectacular events more than on long-term commonplace ones. We need to demand that popular publications find ways to make the implications of complex scientific and quantitative data accessible to nonspecialists. At the same time, we may need to reeducate ourselves so we can better understand such data. Of course, not all decisions can or should be made on a strictly scientific or quantitative basis. Personal factors (peace of mind) and other pressures (satisfying the demands of a family) are often important.

In addition to gathering better information, we need to learn sound processes for evaluating the information we have. We might try to study the system experts use. While most experts work for government or industry, and while they focus not on personally chosen risks but on socially imposed dangers, their system for risk assessment may work at times for us. Their analyses have, in general, three main stages:

- an agreed-upon definition of the hazard;
- an informed estimate–based on the most nearly complete available information–of the level and extent of possible harm associated with the hazard; and
- a careful evaluation of the acceptability of the danger relative to other hazards.

We face many risks, private and public, some (though not all) quantifiable. We need to accept the idea that some of our risks can be reduced, perhaps eliminated–though possibly at great cost–and that others can, perhaps must, be lived with.

Risk-management choices are like some political choices. They are about the competition for limited resources. Any steps to reduce one risk requires the use of some of our resources, taking away from what we have left to deal with others. And risk assessment itself can be a risky matter. Both probabilities and costs may be difficult to estimate, so assessments can easily be wrong.

But experts persist because they believe that–contrary to popular wisdom–limited knowledge about the risks we must take is better than no knowledge at all.

Television, reflecting great public interest and concern, provides extensive coverage of natural disasters with strong visual impact. Yet television is often criticized for the way it presents risks. To date, very few attempts have been undertaken to assess the role of the broadcast media in elaborating and reinforcing dominant public perceptions of risks during disasters.

Risk and Certainty Equivalence Applet

Why are some people willing to buy lottery tickets but, at the same time, insure themselves against theft, death, or property damage? The cost of a lottery ticket is substantially more than the average winnings one can expect to get. If everyone purchased lottery tickets every week over one's entire life, few people would come out ahead. On the other hand, the premium we pay for insurance is substantially greater than the average cost of claims. If everyone carries car insurance, relatively few people will file claims in a given year above the cost of the premiums.

People differ in how much they are willing to take risks and what kind of stakes are worth taking a risk for. A lottery ticket costs only a dollar and has little impact on our lifestyle, while the potential multimillion dollars payoff would impact our lives greatly. Paying $1,000 or $2,000 in insurance premiums for a house or car is costly but perhaps is worthwhile if it excuses us from the unlikely but costly lawsuits resulting from our causing injury.

Economists often express one's willingness to take risks through a utility function of money. To understand the notion of "utility over money," consider the following illustration: A "prize patrol van" pulls up in front of your house. When you open the door, the representative of the state lottery announces that you have won $1 million. As you scream in glee, a decibel meter records the volume (loudness) of your scream. Now imagine the same scenario, only you are told that you have won *$2 million.* Do you scream twice as loud? Probably not. Winning $1 million probably makes you very happy. Winning $2 million also makes you happy, but not *twice* as happy as $1 million. Consider the following utility function.

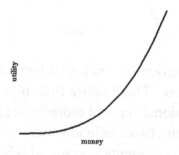

The horizontal axis represents the amount of money a person has, and the vertical axis represents "utility" or how much that money is worth to us. Note that the utility function is convex (loosely, this means that it increases slowly initially and then faster). What does this utility function say about risk? It implies that a person is *risk seeking*, or likely to take gambles (buy lottery tickets or play blackjack at the casino). To see why, consider a poor college student with a $2 bank balance at the end of the month, the day before $100 in rent is due. If the student gave up $1, his condition would be little changed. His "utility" of $1 or of $2 is almost the same, since neither is enough to pay the rent. However, $100 is worth substantially more to the student. Therefore, he may be willing to buy a lottery ticket with one or both of his dollars, thinking, *Even if I lose, I am still in trouble with the landlord, but if I win, I am saved!* The convex utility function represents this: small increases in money above zero wealth have little impact on one's utility, but larger increases make one substantially better off. Now consider this utility function:

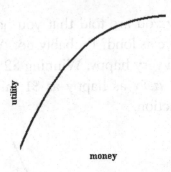

This utility function is concave–it increases quickly initially and then flattens out. This implies that money, initially, is more valuable than additional sums of money once we are already rich (why we don't scream twice as loud when we win twice as much). The utility function represents a person who is *risk averse* or prefers not to take risks. A third type of utility function is this one:

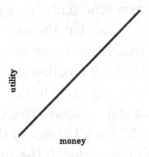

This utility function simply represents that every dollar is worth to us as much as every other. Such a person is deemed *risk neutral.*

> ***A beautiful face will soon age***
> ***A perfect body will certainly change***
> ***But a beautiful soul my dear friends***
> ***Will be beautiful at each and any stage.***
> **-Reza Noubary**

CHAPTER 7

Decision-Making

*The risk of a wrong decision is preferable
to the terror of indecision.*
–Maimonides

T HIS CHAPTER IS concerned
with decision-making and its
risks. The goal is to show how, even when applying the right
mathematics, one may face misleading result or certain difficulties.
We start with some demonstrating examples.

Who Is a Better Investor?

Suppose that an investor picks a hundred stocks, and thirty
of them give a return of over 5 percent in a given period. Let us,
for simplicity, replace the investors by basketball players and the
returns over 5 percent by free-throw percentages.

So, suppose that during the last basketball season, Jim attempted one hundred free throws in the first half of the season and made thirty. So his free-throw percentage was $30/100 = 0.300$. He also attempted twenty in the second half of the season and made eight ($8/20 = 0.400$). His statistics were better than Curt's, $5/20$ (0.250) for the first half of the season and $35/100$ (0.350) for the second half of the season. For the season, however, Curt's free-throw percentage, $40/120 = 0.333$, was higher than Jim's, $38/120 = 0.317$. Using this information, which would you choose as a "better" free-throw shooter? The following example makes this even clearer.

Which Treatment Should I Choose?

Consider a disease for which there are two methods of treatment, A and B, and it is up to the patients to choose one. Suppose that in the past, out of 100 male patients who chose A, 20 recovered (recovery rate $= 20/100 = 0.20$). Also, out of 210 male patients who chose B, 50 recovered (recovery rate $= 50/210 = 0.24$). These rates suggest that a male patient should prefer treatment B to A (24 percent versus 20 percent). Suppose also that, so far, out of 60 female patients who chose A, 40 recovered (recovery rate $= 40/60 = 0.67$). Also, out of 20 female patients who chose B, 15 recovered (recovery rate $= 15/20 = 0.75$). These rates suggest that a female patient should prefer treatment B to A (75 percent versus 67 percent) as well.

Now, combining the data, we see that the total number of people (regardless of their gender) who chose A was 160, of which 60 recovered (recovery rate $= 60/160 = 0.38$). Also, the total number of people (regardless of their gender) who chose B was 230, of which 65 recovered (recovery rate $= 65/230 = 0.28$). These rates suggest that it is wise to use or choose treatment A over B (38 percent versus 28 percent). This, unlike expectation, contradicts our earlier conclusion based on patient's gender.

The lesson here is that with only two numbers, as in this example, drawing a conclusion is not always straightforward. From a mathematical point of view, the conclusion based on details (higher dimensions/gender) should be preferred. In fact, breaking the data gives a two-dimensional view of it, which is better than a one-dimensional view of it based on collapsed data.

The situation described is an example of what is known as Simpson's paradox. The paradox occurs because collapsing the data can lead to an inappropriate weighting of the different populations.

Pay Discrimination

Consider a business with one hundred employees, fifty "type A" and fifty "type B," with comparable duties and average pay of $16,000 (total pay = $800,000) and $14,000 (total pay = $700,000) per person, per year, respectively. Here, the average pay to type A employees is more than that of type B employees, suggesting a pay discrimination in favor of type A employees. But by considering the details, one finds that the data contains other relevant information. The length of employment also factors into pay rates.

Suppose that ten of the type A employees have been working for that business for less than five years with an average pay of $10,000 (total pay = $100,000) and forty for more than five years with an average pay of $17,500 (total pay = $700,000); $100,000 + $700,000 = $800,000.

In type B group, forty employees have been working there for less than five years with average pay of $12,500 (total pay = $500,000) and ten for more than five years with average pay of $20,000 (total pay = $200,000); $500,000 + $200,000 = $700,000.

Now, comparison suggests that type B employees' average pay is higher than that of type A employees in both categories, suggesting a pay discrimination in favor of type B employees. This again contradicts our earlier conclusion.

These examples demonstrate how different analyses of the same data may lead to apparently conflicting results and sometimes incorrect decisions. So here is the lesson: making decisions based on averages is not straightforward—that is, one must be careful in reporting averages unless the groups under consideration are homogeneous. The problem gets even more complicated (and interesting) if we consider more than one classifier (e.g., education, experience).

In sum, these examples demonstrate how different analyses of the same data may lead to apparently conflicting results and sometimes incorrect decisions.

Decisions Based on Statistical Testing

Every day we hear lots of claims by individuals and groups, much of which require clarification or investigation. A tool used for this is statistical testing, which is an integral part of inferential statistics. The problems considered under this title include testing a hypothesis about the unknown parameters of a distribution/population, testing claims including scientific ones. The decision (rejecting or not rejecting the hypothesis) is made by using the data (sample) as evidence and by considering how much error one is willing to tolerate when making such decision. Testing hypotheses also includes topics such as establishing relationships among variables of interest as well as detecting differences among the groups. It is important to remember that the major part of science and scientific investigations are based on certain accepted or agreed-upon theories or hypotheses. Examples include evolution theory, big bang theory, tectonic theory, etc. These theories are used as long as no plausible alternative supported by data is introduced.

Let us start with a real-life example. It is known that companies frequently conduct surveys to help them make decisions concerning the effectiveness of their advertising. Suppose that a company that produces running shoes and has 10 percent of the market share launches a new advertising campaign. At completion, the company wants to know if the campaign was successful in raising its market share. A random sample of purchasers of this shoes will provide evidence upon which a conclusion may be drawn (an answer may be found). Here, the company should test the null hypothesis $p=0.10$ (no change/not effective) versus $p>0.10$ (increase/effective), where p presents the market share after the campaign. These values present two alternative hypotheses—namely, advertising was not effective and advertising was effective.

Decision Process

To understand the decision-making process, suppose that we are given a coin and asked to decide whether or not it is a fair coin. One reasonable approach is to flip the coin twenty times, say, and decide based on the outcome. Note that the number of times we may flip the coin is a decision in itself. However, we are not going to discuss this problem at this point. Next, we need to define our decision criterion. Considering the fact that, theoretically, we expect to observe ten heads, we may call the coin fair if we observe between seven (10-3) and thirteen (10+3) heads, or be a little bit more conservative and call it fair if between eight and twelve heads are observed. The question that arises at this point is, how do we decide which criterion is "better"? To find an answer, let us assume that the coin is fair. If this is the case, we ask, what is the probability of making a wrong decision and declaring the coin "not fair"? Let us call this error type I error. Using the binomial distribution, this probability can be calculated as

$$1-P(7 \leq X \leq 13) = 1 - \sum_{x=7}^{13}\binom{20}{x}(1/2)^x = 0.116$$

for the first decision criterion, and as

$$1-P(8 \leq X \leq 12) = 1 - \sum_{x=8}^{12}\binom{20}{x}(1/2)^{20} = 0.264$$

for the second one. This shows that by requiring stronger evidence, one would face a higher risk of rejecting a true hypothesis (coin is fair). Stating this in a reverse order, if one can tolerate a higher risk of rejecting a true hypothesis, then a deviation of observed from the expected outcome (in this case ten heads for a fair coin) will provide stronger evidence against the null hypothesis (coin is fair).

On the other hand, suppose that the coin is not fair and the chance of observing a head is, for example, 60 percent. Then the probability of declaring the coin fair when it is not is

$$P(7 \le X \le 13) = 1 - \sum_{x=7}^{13} \binom{20}{x}(0.6)^x(0.4)^{20-x} = 0.744$$

for the first criterion and

$$P(8 \le X \le 12) = \sum_{x=8}^{12} \binom{20}{x}(0.6)^x(0.4)^{20-x} = 0.563$$

for the second one. Let us call this error type II error. If, rather than 60 percent, probability of observing a head is 80 percent, then respective probabilities are

$$P(7 \le X \le 13) = \sum_{x=7}^{13} \binom{20}{x}(0.6)^x(0.4)^{20-x} = 0.087$$

$$P(8 \le X \le 12) = \sum_{x=8}^{12} \binom{20}{x}(0.8)^x(0.2)^{20-x} = 0.032$$

So what is the conclusion? Well, we have to think about two factors: the type of the error that is important and the level of the error that we can tolerate. The next step is to use these to define a decision criterion. For example, rejecting a null hypothesis that is true may be considered more serious than accepting an alternative hypothesis (not fair) that is false–that is, type I error is more important than type II error. We will elaborate on this point in the later sections. For example, a court may declare a person guilty when in fact the person is innocent is more important than declaring a person innocent when the person is guilty. Whereas when testing for HIV, false negative is more important than false positive. Let us look at the first example in more detail. Also, in preventive medicine, declaring a person diabetic while he or she is not is less risky than the opposite and, in the long run, less costly. The same is true when testing for HIV since a false negative leads to a more severe consequence, through infecting other people.

We end this part by providing a slightly different clarification using the so-called rare event rule, which constitutes a foundation for hypothesis testing. According to this rule, if under a given hypothesis the probability of a particular observed event is "small" (for example, less than 5 percent), then we may conclude that the hypothesis is probably not correct because unlikely outcomes are not expected to happen. When they do, there are two possibilities: either a very rare thing happened or our assumption/hypothesis is false. For example, if we roll a fair die one hundred times, we expect to observe (approximately) an even number fifty times. For a fair die, the chance of observing an even number thirty, forty, and forty-seven times are, respectively, 0.00004, 0.028, and 0.31. We see that, except for the last case, the results are significant (less than 0.05)–that is, events are rare. Thus, if any of the first two events occur, we can safely conclude (at 5 percent level) that the die is biased. The above three probabilities are called p-value (probable value), and 5 percent is known as level of significance. In sum, the statistical method for testing hypotheses considers small probability of occurrence as a strong evidence in favor of the alternative, simply because the chance of such an outcome is very small. In other words, the p-value (probable value) is small and we do not expect this event to occur. So the occurrence of such (rare) event provides evidence against the null hypothesis.

Right, Wrong, or Maybe Both?

In the coin example above, we had three p-values, 0.00004, 0.028, and 0.31, and the level of significance was 0.05. So if we observe thirty or forty heads, we reject the null hypotheses that the coin is fair and not when we observe forty-seven heads.

Now, suppose that we choose a smaller level of significance such as 0.025–that is, we can only tolerate 2.5 percent chance of rejecting the true null hypotheses (fair coin). This means that we need a stronger evidence against the null hypotheses to reject it. In the case of our example, the number of heads observed should deviate a great deal from the expected fifty. Here, the p-value for forty is 0.028, which is not less than the level of significance 0.025, so we do not reject the null hypothesis; whereas with 0.05, we reject. This has a great implication on life and decision-making. Things could be rejected or not rejected. It all depends on sample size chosen level of significance.

Statistical Hypothesis

A statistical test is a method to decide whether or not to reject a statistical hypothesis by inspecting a sample. To clarify, suppose that in a community, the distribution of income is normal with mean $50,000 and standard deviation $10,000. You meet someone whose income is $84,000. You wonder whether this person is a member of this community. This could be the null hypothesis. The hypothesis that the person is not a member of this population could be the alternative hypothesis. Let X denote the income of a randomly selected member of this community. To test the hypothesis of interest, we may ask the following question: what is the probability that a randomly selected member of this community makes $84,000 or more? This is easy to answer. We have

$$P(X > 84,000) = P(z > \frac{84,000-50,000}{10,000}) = P(z > 3.4) = 1 - 0.9997 = 0.0003.$$

This mean that only three people out of ten thousand have this type of income. Thus, either this person is one of the very few people in this community, or, else, he or she is not a member of this community. But which of these two possibilities is more likely? Asking the question differently, our observation (person making $84,000 or more) provides stronger evidence in favor of which hypothesis, "is" or "is not" a member of this community? The statistical method for testing hypotheses considers the above probability as a strong evidence in favor of "is not" a member of this community, simply because the chance of such an event occurring is very small. In other words, the p-value (probable value) is small and we do not expect this event to occur. So the occurrence of such (rare) event provides evidence against the null hypothesis.

Coin Example

A coin is flipped ten times and eight heads were observed. The chance of such event is less than 5 percent (in fact, 0.044 using binomial distribution with $p=1/2$) for a fair coin. This, therefore, provides evidence against the null hypothesis that the coin is fair. Again, this is because the probability of such an event happening (p-value) is very small (a rare event). For further discussion concerning p-values, see section 6.5.

More about Errors

Let us consider what happens in a court of law, where we have "not guilty" and "guilty." As discussed, it is possible to declare the person guilty when he (she) is, in fact, not guilty (type I error) or the opposite (type II error)—that is, declaring the person not guilty when he (she) is guilty. The probability of a type I error is the probability of rejecting a null hypothesis when it is true. It is called the significance level of the test or level of significance. In most cases, a type I error is fixed in advance. However, there are cases where a type II error is of prime importance and may be fixed in advance. For example, for most courts, the type I error (declaring an innocent person guilty) is of prime concern. But in some medical tests, false negative (type II error) may be more critical than false positive (type I error). For example, in preventive medicine, declaring a person diabetic while he or she is not is less risky than the opposite and, in the long run, less costly. The same is true when testing for HIV since a false negative leads to a more severe consequence, through infecting other people.

Summary

A statistical hypothesis is a claim or a conjecture about a population parameter. This conjecture may or may not be true.

The null hypothesis, symbolized by H_0, is a statistical hypothesis that states that there is no difference between a parameter and a specific value, or states that there is no difference between two parameters. *Ho* is initially assumed to be true.

The alternative hypothesis, symbolized by H_1, is a statistical hypothesis that states that there is a difference between a parameter and a specific value, or states that there is a difference between two parameters. It usually contains a statement of inequality such as $>$, $<$, or $\neq$.

Type I error occurs if a null hypothesis is rejected when it is actually true.

Type II error occurs if a null hypothesis is not rejected when it is actually false.

Description of Possible Errors

Decision	Truth	
	H_0	H_1
Reject H_0	Type I Error	Correct Decision
Do Not Reject H_0	Correct Decision	Type II Error

Note that decreasing the probability of making one type of error will lead to an increase in the probability of making the other error.

One question of those who are just beginning to learn statistics sometimes ask is whether the type I or type II error should be identified with the diagnosis of illness or well-being. They also often ask which error is the more important. We already discussed this briefly. Consider a hypothetical scenario like this one: suppose the local government imposes mandatory HIV blood tests on all citizens. Assume the tests have a .01 false positive rate and a .01 false negative rate. Unknown to the testers, 500 out of 170,000 citizens are HIV-positive. As a result of this, we should expect 5 false negatives and 1,695 false positives out of 170,000 tests.

Which is the more serious error? Is it 5 undetected HIV carriers or 1,695 people who are falsely believed to be HIV-positive? This leads to catch on to the point that the rarity of a disorder or diseases can not only make the accuracy of a test problematic but also alter our perceptions of which error is the worse one. We also start to see some of the difficulties that arise from using imperfect diagnostic tests on nonclinical populations. Note that under different circumstances, one error may be perceived as more serious than the other, and so we need to worry about both.

z-test. The z-test, as is clear from the name, is based on the normal distribution. To see how it is applied, we summarize the procedure described in section 6.2 as follows:

1. Assume that X is distributed as $N(\mu,\sigma^2)$ and σ^2 is known;
2. Null hypothesis to be tested is $H_0 : \mu = \mu_0$. We may introduce $H_1 : \mu \neq \mu_0$ as an alternative hypothesis;
3. Test quantity is (evidence) $T(x_1,\ldots,x_{N)}) = \dfrac{\bar{x} - \mu_0}{\sigma / \sqrt{N}}$;
4. Test statistic is $N(0, 1)$. This is the distribution of $\dfrac{\bar{x} - \mu_0}{\sigma / \sqrt{N}}$;
5. Critical value is $z_{1-\alpha/2}$ corresponding to the level α;

6. Decision rule is to reject H_0, if $|T| > z_{1-\alpha/2}$. If $|T| \le z_{1-\alpha/2}$, then H_0 cannot be rejected.

Note that by applying the above steps, we have controlled the type I error—namely, the probability of rejecting H_0 when it is actually true. However, on the other hand, we did not pay attention to the type II error—namely, the probability of not rejecting H_0 despite being wrong. The type II error may well happen in 50 percent of the cases. That is the reason to formulate H_0 "negatively." We would like to verify at a high significance level that H_0 is wrong. If we reject H_0 when H_0 is actually true, we have committed an error less than or equal to α. If, on the other hand, we want to verify that H_0 is true and that the test does not reject H_0, then there may be a 50 percent probability that H_0 is wrong. The "proof" or "disproof" of H_0, then, is just like flipping a coin. As mentioned earlier, one interesting question that arises in testing a hypothesis is, which error is more important and should be controlled? For example, most courts try to avoid the type I error, declaring an innocent person guilty. However, in other cases such medical tests where we could have false positives or false negatives, situations may be different. As an example, when testing HIV, clearly a false negative is more important and should be avoided, if possible.

Note: If distribution of X is not normal but N is sufficiently large, then according to the Central Limit Theorem, the distribution of $\overline{X}$ is approximately normal and z-test may still be applied. Moreover, for large samples, we can replace sample standard deviation s for σ (when it is unknown) and follow the steps described above. This, of course, will provide us with an approximate procedure.

Example. Data was collected on the golf ball driving distances by professional golfers in a recent tournament. The data on the first drives of thirty-six randomly selected golfers yielded a sample mean distance of 247.5 yards. Assuming the standard deviation of the driving distances on the first drive is ten yards, carry out a z-test on

$$H_0 : \mu = 250 \text{ versus } H_1 : \mu \neq 250 \text{ at the}$$
10 percent significance level.

Solution: Let X denote the driving distance of a randomly selected golf ball. Assumption: X is normal $N(\mu, 10^2)$. We have

$$T = z = \frac{247.5 - 250}{10 / \sqrt{36}} = -1.5.$$

The critical value in this case is $z_{0.95} = 1.65$. Since $|T| < 1.65$, H_0 cannot be rejected–that is, although we have some evidence against H_0, it is not "strong" enough (at 10 percent significance level) to justify its rejection. Note that the same evidence may be considered "strong" for a different (in this case higher) level of significance–that is, if we can tolerate a larger error, we may consider the evidence strong (in that level) and reject the null hypothesis.

Example: In recent years, college basketball in United States has become increasingly popular since it creates types of excitement and competition that is not present in professional sports. One rule of these games is the thirty-five-second shot clock, which fans have some concern about. They think that it slows the games and makes them less exciting. However, NCAA officials claim that this rule has not affected the game in any way fans think. Data collected shows that the average time that a college team takes to set up a shot was twenty seconds in the past and is still twenty seconds. To test this, a random sample of games played using the clock is selected. The average time taken to set up a shot based on 625 opportunities is found to be 19.1 with s=7.2. Does data support the claim made by the officials? Use $\alpha = .05$.

Solution: Since we have a large sample, the hypotheses

$$H_0 : \mu = 20 \text{ vs } H_1 : \mu < 20$$

may be tested using

$$T = z = \frac{19.1 - 20}{7.2 / \sqrt{625}} = -3.125.$$

The rejection region is $T < -1.654$. Thus, we have sufficient evidence to conclude that the true average time taken to set up a shot is even less than twenty seconds claimed by officials.

Although other types of alternatives and tests for proportions are not yet discussed, here we present figure 6.3.1, which illustrates the way they work. We also present a different summary of the terms and procedure used for hypothesis testing below.

Problem: A new diet and exercise program claims that participants will lose, on average, at least 8 pounds during the first week of the program. A random sample of forty participants in the program showed a sample mean weight loss of 7 pounds with a sample standard deviation of 3.2 pounds. Would data support the claim made by the diet program at 5 percent and 2 percent significance levels?

Solution: X = weight loss of a randomly selected participant and $\mu = E(X) =$ the average weight loss of participants. Since n>30, according to the central limit theorem, distribution of sample mean is approximately normal. We want to test $H_0 : \mu = 8$ vs. $H_1 : \mu < 8$ (one-tailed). Since the test stat $T = \sqrt{40}(7-8)/3.2 = -1.98$ and the p-value = $P(z \le -1.98) = 0.0239 < 0.05$, we reject H_0 and conclude that there is sufficient evidence to indicate that the true mean weight loss under this program is less than 8 at 5 percent level of significance. Now, if we change the level of significance to

0.02, then since $0.0239 > 0.02$, we do not reject H_0 and conclude that there is no sufficient evidence to indicate that the true mean weight loss under this program is less than 8 at 2 percent level of significance. This indicates that we may reject a null hypothesis in one level and do not reject in another level.

CHAPTER 8

Risk and Faith

If a man is not willing to take risk for his
opinions, either his opinions are no good
or he is no good.
–Ezra Pound

A FRIEND OF MINE who, as he says, does not believe in God had to call 911 for his sick mom. We were all very nervous and could not wait for the ambulance to arrive. I looked at him and, to my surprise, noticed that he was quietly saying, "God, please help them to get here soon." I once more realized how weak and fragile we are. Fortunately, all went well. When I returned home, I was thinking about many critical questions: Am I accountable for all the choices that had been made for me such as my birthplace, country, time, gender, parents, religion, culture, etc.? Should I choose faith and accept their answers?

Science and Faith

The history of science includes countless hard-fought battles with traditional certainties. A classic example is the fierce conflicts of belief associated with Creationism's rejection of the theory of evolution. It seems that wherever modern science goes too far and establishes itself as a kind of alternative to religion, it produces a backlash. The root of conflicts is that in the world of religion, miracles are certainly possible; in the world of science, they are not. Additionally, brain as a belief engine searches for understanding and drives us at least to believe, even if we do not actually know about God, astrology, extraterrestrials, or string theory.

All this takes place because of our ability to link cause and effect. Once we understood the connection between cause and effect, we could not stop searching for reasons why the world is as it is. Once we realized that we only have to rub two sticks together vigorously to create fire, we wanted to know the causes of other things such as disease and death. But it proved impossible to apply the principle of "no effect without a cause" to strokes of fate of this kind without resorting to the supernatural. When our ancestors reached the limits of their understanding, they almost inevitably came to the conclusion that an invisible God must be responsible—a solution that was so compelling, and has remained so for many to this day, that a lot of societies developed it quite independently. Additionally, this line of thought answers our major questions simultaneously and, at the same time, gives us something to hold on to.

After all, we are human, weak, fragile, and not equipped with tools or intelligence to uncover the secrets of this complex world. Faced with uncertainties, we feel lost and often panic. To comfort ourselves, some of us ignore them, some deny them, some avoid them, and some try to understand them. Some of us even find believing in fate or even conspiracy more comforting than admitting ignorance.

Should I Believe in God?

For the reasons mentioned, most people believe in some form of supernatural power. Why? Consider the Blaise Pascal's philosophical argument regarding God. Think about two events: the event that God exists and the event that a person is a believer that God exists. Suppose that p is a number between zero and one, the probability that God exists. Each person has options of becoming a believer or nonbeliever. This leads to four disjoint (mutually exclusive) possibilities. God exists and person is believer, God exists and person is nonbeliever, God does not exist and person is believer, God does not exist and person is nonbeliever. Suppose that the payoffs of these options can be expressed in dollars as follows:

	God Exists	God Does Not Exist
Believer	100,000	-10,000
Nonbeliever	-100,000,000,000	100,000

To analyze this further, let us first use an example to explain what the expected value or expected return is. Think about a game where you flip a fair coin and win a dollar if head comes up and lose a dollar if tail comes up. If you play this game for a few hours, you expect to break even–that is, expect neither win nor lose. This because you expect to win half of the times and lose half of the times. Mathematically, this is calculated as $1/2(+1)+1/2(-1)=0$.

Using the numbers in the above table, the expected payoffs are $10^5p-10^4(1-p)$ for a believer and $-10^{10}p+10^5(1-p)$ for a nonbeliever. These two values are equal if $p=1/90911=0.000109$. Thus, the expected payoff is higher for believers if the chance that God exists is bigger than just 0.000109. In fact, Pascal's original argument uses infinite rather than -1010 and concludes that believing in God results in better payoff for any value of p, no matter how small.

Discussion

Many scientists believe that our brains search to understand the world around us and, in the process, drive us to believe in God, astrology, extraterrestrials, or string theory. This happens because of our ability to link cause and effect. Once we understood the connection between cause and effect, we could not stop searching for reasons why the world is as it is. We wanted to know the cause of things such as disease and death, but we also found it impossible to apply the principle of "no effect without a cause" to strokes of fate without resorting to the supernatural. When our ancestors reached the limits of their understanding, they concluded that an invisible God must be responsible, a solution that was so compelling–and still is–that a lot of societies developed it quite independently. This line of thought answers our major questions and gives us something to hold on to. After all, humans are weak, fragile, and not equipped with tools or intelligence to uncover the major secrets of this complex world.

Some scientists believe that when faced with uncertainties, we feel lost and tend to panic. To comfort ourselves, we either ignore, deny, avoid, or try to understand those uncertainties. Some of us find believing in fate or even conspiracy more comforting than admitting ignorance.

How Many Gods? Non-Euclidean Geometries/ Statistical Physics

One frequently cited empirical discovery is the general theory of relativity and its adoption to non-Euclidean geometry. Recall that Euclid developed the idea of geometry around 300 BC. In his book, he starts with five main postulates, axioms, or assumptions from which he drove theorems of geometry. The postulates were as follows:

1. Given two points, there is a straight line that joins them.
2. A straight-line segment can be prolonged indefinitely.

3. A circle can be constructed when a point for its center and a distance for its radius are given.
4. All right angles are equal.
5. From a point outside a line, only one line can be a parallel to it.

The last postulate, number 5, is more complicated than the other four. Over the years, many mathematicians have tried to derive it from the first four, but nobody has come up with a proof.

Details

Geometry is the realm of mathematics in which we talk about things such as points, lines, angles, triangles, circles, squares, and other shapes, as well as the properties and relationships between all these things. For centuries, it was widely believed that the universe worked according to the principles of Euclidean geometry where parallel lines never crossed, and this was the only kind of geometry taught in schools. In fact, for a long time, this was my favorite joke: "Parallel lines have a lot in common. It's unfortunate that they can never meet."

The early nineteenth century would finally witness decisive steps in the creation of non-Euclidean geometry. Non-Euclidean geometry arises when the parallel postulate is replaced with an alternative one. Doing so, one obtains hyperbolic geometry and elliptic geometry, the traditional non-Euclidean geometries. The essential difference between the geometries is the nature of parallel lines. Euclid's fifth postulate, the parallel postulate, states that within a two-dimensional plane, for any given line l and a point A, which is not on l, there is exactly one line through A that does not intersect l. In hyperbolic geometry, by contrast, there are an infinite number of lines through A not intersecting l, and in elliptic geometry, any line through A intersects l.

Another way to describe the differences between these geometries is to consider two straight lines, both perpendicular to a third line and indefinitely extended in a two-dimensional plane:

- In Euclidean geometry, the lines remain at a constant distance from each other and are known as parallels. This means that a line drawn perpendicular to one line at any point will intersect the other line and the length of the line segment joining the points of intersection remains constant.
- In hyperbolic geometry, they curve away from each other, increasing in distance as one moves farther from the points of intersection with the common perpendicular.
- In elliptic geometry, the lines curve toward each other and intersect.

In my view, the story of geometries somehow relates to the story of three statistics regarding particles in physics: Fermi-Dirac, Bose-Einstein, and Maxwell-Boltzmann. If we have indistinguishable particles, we apply Fermi-Dirac statistics. To identical and indistinguishable particles, we apply Bose-Einstein statistics. And to distinguishable classical particles, we apply Maxwell-Boltzmann statistics.

Finally, based on our knowledge about geometry and the fact that none of them—Euclidian, hyperbolic, or elliptic—are incorrect, we could infer that assuming no god, or one god, or many gods as our postulate would not necessarily lead to any inconsistency or contradiction.

References

"Geometry." *Wikipedia.* Wikimedia Foundation. 2 May 2019. en.wikipedia.org/wiki/Geometry.

Humboldt Network. www.humboldt-foundation.de/web/humboldt-network.html.

"Non-Euclidean Geometry." *Wikipedia.* Wikimedia Foundation. 6 May 2019. en.wikipedia.org/wiki/Non-Euclidean geometry.

CHAPTER 9

Probability and Uncertainty

Life is the school of probability- Walter Bagehot

WE MAY NOT actually compute probabilities in our daily lives, but many of our daily decisions rely on our intuition about probability. Although this has been cultivated in us for many years, it might come as a surprise how this intuition can be remarkably unreliable.

Luck: An Invisible Player in the Game of Life

People who get cancer often view it in terms of cause and effect: How did this happen, and more importantly, how could I have avoided it? This is understandable based on what we hear and read. However, a recent study by oncologist Bert Vogelstein and bio-mathematician Cristian Tomasetti, published in the journal

Science (Jan 2015: Vol. 347, Issue 6217, pp. 78-81), concluded that most cancers are not influenced by lifestyle or environment, but that cancer is the result of bad luck. In "Biological Bad Luck Blamed in Two-Thirds of Cancer Cases," which summarizes Vogelstein and Tomasetti's study results, Will Dunham notes that two-thirds of cancer incidents can be blamed on random mutations accumulating in various parts of the body during ordinary cell division rather than on heredity or risky habits.

Vogelstein and Tomasetti looked at thirty-one cancer types and found that twenty-two of them, including leukemia and pancreatic, bone, testicular, ovarian, and brain cancers, could be explained largely by these random mutations—essentially biological bad luck. The other nine types, including colorectal, skin (known as basal cell carcinoma), and smoking-related lung cancers, were more heavily influenced by heredity and environmental factors such as risky behavior or exposure to carcinogens. Overall, they attributed 65 percent of cancer incidents to random mutations in genes that can drive cancer growth. They examined the extent to which stem cell divisions in healthy cells—and the random mutations, or "bad luck" accumulations—drive cancer in different tissues.

What else can be said about bad luck? This question was highlighted when researchers tried to sort out environmental versus inherited causes of cancer. Their effort implied that cancer is harder to prevent than previously hoped and that early detection is underappreciated. This sparked controversy and confusion among people who had believed that, though it's a complex topic to put numbers on, many cancers could be prevented largely through lifestyle changes. This concern has also been addressed by Michael Walsh in "Reports That Cancer Is 'Mainly Bad Luck' Make a Complicated Story a Bit Too Simple."

To reexamine the luck theory, Vogelstein and Tomasetti did further study and again published their findings in *Science*, and again the media picked up their results. The second study addressed some of the original concerns about approach by boosting sample size and adding data from more countries and cancer types. The results supported their original conclusions, adding further evidence to their theory that cancer risk is most strongly associated with how quickly specialized cells, called stem cells, replicate. They predict that mutations in two out of three cancers are due to bad luck.

Although *Science* is a well-respected journal, some people still wonder whether this message does more to confuse than inform. According to Walsh, one of the biggest problems is that trying to get a clear answer to the cause-and-effect question from this analysis might oversimplify things. When cancer develops, many other real-world factors are at play, and it's critical to understand how those factors affect both the progression and the prevention of cancer. Mutations that can lead to cancer can be caused by environmental and lifestyle factors, such as cigarette smoke, but they can also be inherited or happen by chance. These chance mutations occur when errors are made as a cell copies and divides its DNA. Vogelstein and Tomasetti believe that this third factor is a major player in cancer cases, and in their model, it accounts for around two in three mutations in cancer. This, they say, is the bad luck factor—and they want that to be acknowledged more often.

Final Remarks

We learn about the world around us by collecting and analyzing data. Unfortunately, the world around us is so complex— and our knowledge so limited—that our views and understanding of data hardly match reality. For example, scientific modeling takes the data as a message and tries to model them using the basic decomposition.

Datum = Smooth Part (Trend) + Rough Part
  = Systematic (Deterministic) Part + Random Part
  = Signal + Noise

Mathematics is used to model the systematic part; whereas, probability and statistics are used to analyze the random part. Models developed are often judged on their signal-to-noise ratio. When the ratio is bigger than one, the smooth part is dominant. This allows us to use the smooth part for prediction and the rough part for assessing our model and its precision. In the opposite case, where the random part is dominant, we often do not know much about the phenomenon under consideration—and so we attribute them to luck or chance.

References

Dunham, Will. 2015. "Biological Bad Luck Blamed in Two-Thirds of Cancer Cases." *Daily Star.* http://www.dailystar.com.lb/Life/Health/2015/Jan-04/283044-biological-bad-luck-blamed-in-two-thirds-of-cancer-cases.ashx.

I-Digest: Indonesia Digest. www.indonesia-digest.net.

Walsh, Michael. "Reports That Cancer Is 'Mainly Bad Luck' Make a Complicated Story a Bit Too Simple." *Cancer Research UK.* https://scienceblog.cancerresearchuk.org/2017/03/24/reports-that-cancer-is-mainly-bad-luck-make-a-complicated-story-a-bit-too-simple/.

Probability Theory

Probability theory investigates (systematically) the laws concerning phenomena influenced by chance. It is the branch of science that studies methods for making inference for situations that involve uncertainty. The theory starts with introducing a measure for degree of certainty (uncertainty) about outcomes. In recent decades, probability theory has become exceedingly important. One reason for this stems from the fact that almost everything in real life involves some degree of uncertainty. This is even true for science itself since, except for mathematics, which is manmade, other disciplines are empirical based and as such there is a margin of error. Well-known examples are evolution, big bang, tectonic theories, etc.

Probability theory also provides tools for making quantitative statements about uncertainty and allows one to draw conclusions from such statements using certain rules. It is (as opposed to the word "probability") a mathematical theory (a model of reality) that enables us to calculate the likelihood of outcomes. Areas of applications include reliability and risk analysis, quality control, business, social sciences, medical sciences, insurance, etc.

In 1620, Francis Bacon argued that learning about the world could only take place by observation and induction. With the rapid expansion of experimental sciences, scientists began to make use of the probability theory. For example, in a monumental piece of experimental research between 1856 and 1863, Mendel laid the statistical laws of genetics. Without any precedent, Mendel perceived that the genetic mechanism operated like a random device.

What Is Probability

Probability is not just a vehicle but a standalone field with a large number of unanswered questions. It is, in a way, a human response to the seeming lack of complete determinism in the world where there is uncertainty. Probability may even be seen as human reaction to fear and something that guides much of decisions we make.

Most people use probability for the exact reason that the field was originally created: to handle the concern of the unknown. It is used to provide a way to quantify the visceral fear of unpredictability. Like any other field of mathematics, probability is a human way to shape the uncertain world into something we can understand or feel comfortable about.

For example, in medical field with each test, there are four possibilities. The patient has the disease and test result is positive, the patient has the disease and the test results is negative, the patient does not have the disease and the test result is positive, and the patient does not have the disease and the test results is negative. The probability of each of these events varies depending on how accurate the test is and how common the disease is. Think about diseases such as neurofibromatosis, Huntington's, Tay-Sachs, and cystic fibrosis, which are all genetic diseases that can be passed from parents to their children. A couple with a family history of these diseases may consult to figure out the probability of the child inheriting a genetic disease. Finally, probability is used in all other branches of science. For instance, probability is used in the discipline of engineering for various reasons ranging from quality control to quality assurance.

Uncertainty Makes the Life Exciting

Although uncertainty may seem a problem, it could make life interesting and exciting. Indeed, the world would be a dull place if things were completely predictable. Among many real-life examples of this, we find sports exciting because the outcomes are not completely predictable. Of course, uncertainty can also cause grief and suffering.

Let me start with an everyday example. Probability theory is also a useful tool for making inference. Suppose that as you come out of your apartment, you notice a man looking inside your car through the window. You will probably infer that the man is trying to rob your car. How did you arrive at that conclusion? You considered the possible set of circumstances that might have produced the sample of event that you observed. Was the man trying to rob your car? Was the man interested in a car like yours and wanted to see how the interior of the car looks? Was the man trying to figure out if this was his friend's car? Of these and many other possibilities for the event you observed, you picked the outcome that you thought was most probable—that is, you based your inference on your assessment of a set of probabilities.

Probability Modeling

Mathematical modeling is a very fertile field in modern scientific investigations. Its goals is to analyze and translate real-life situations into scientific terms. No one can analyze the real world (by definition of analyze). One can only analyze a picture (conceptual model) of the world.

Approaches to modeling and data analysis have changed a great deal in recent decades. Probability is one important concept used in such approaches to mathematical modeling. By using probability, data is decomposed into a smooth part and a rough part or systematic part and random part (signal and noise), and models are developed for each part separately.

It is the attention given to the second component that perhaps most distinguishes probabilistic approaches to the modeling. In fact, mathematics mostly deals with the analysis and modeling of the systematic part. In the modern approaches, however, models are judged based on their signal-to-noise ratio. This ratio measures the contribution of the systematic part relative to the random part. Often, a large value of this ratio occurs if the context of data is well understood. When this happens, usually only the systematic part is used as a model to describe the situation and, more importantly, to make predictions. The random part is then used to evaluate the model and to produce bounds on prediction errors.

In addition to this, today, the ideas of randomness are central to much of the modern scientific disciplines. Clearly, the real world can only be analyzed and explained scientifically through physical sciences. However, many scientists expressed their frustration when attempting to do this utilizing classical method based on determinism. Historically, it was natural for physicists to be interested, at first, in the macroscopic world that surrounds us. To make quantitative predictions about it, they devised deterministic models, which perform impeccably. Such are the origins of mechanics, of thermodynamics, of optics, of electromagnetism, and of relativity. These theories remain valid in the domains for which they were designed, and they continue in a state of vigorous development. But as regards to fundamentals, physics today has its cutting edge on the microscopic level, where progress is achieved by means of probabilistic models, models that allow precise quantitative predictions for random phenomena. In short, chance is inherent in the basic nature of microscopic processes, reducing determinism to a mere consequence of chance regarding mean values that are on the macroscopic level.

Our Intuition: Birthday Problem

Birthday is an important date for most people, and as such, they show interest in problems related to it. Think, for example, about birthday matches in a classroom or church. There are 365 days in a year, and so we need to assemble 366 people in a room to guarantee at least one birthday match. That is straightforward. But how many people do we need in a room to guarantee a 50 percent chance of at least one birthday match? I ask you to guess a number. When I ask this question, usually the numbers vary a great deal. The answer, to most people's surprise, is 23.

Here are further facts about birthday matches that often surprise people:

- With only 41 people, the chance of at least one common birthday is more than 90 percent.
- With 88 people, the chance of at least three common birthdays is more than 50 percent.

Here are more interesting facts:

- If we want a 50 percent chance of finding two people born within one day of each other, we only need 14 people.
- If we are looking for birthdays a week apart, the magic number is 7.

- If we are looking for a 50 percent chance of finding someone having a specific person's exact birthday, we need 253 people.

I often ask students to calculate the probability that two randomly selected individuals have birthdays, for example, on May 25 and October 13. I usually choose two students in the class and use their birthdays. The answer (after providing hints) they should find or suggest is $(1/365)2$. We then change the dates to, say, June 9 and June 9 (again, we choose someone's birthday in the class). Not very sure they eventually produce the same answer. In both cases, the probability is less than $1/133,000$. I ask them if they consider this a small probability. Most students answer yes. We ask which of these two events they find more surprising. All of them, of course, refer to the latter.

This reveals that, although these two events are equally likely, the latter creates a great deal of surprise and therefore may be classified as a coincidence. We ask if they are convinced that the two main components of a coincidence are probability and the degree of surprise.

Why do we classify some events as coincidences? According to John Allen Paulos, human beings are "pattern-seeking animals." It might just be part of our biology that conspires to make coincidences more meaningful than they really are. A sport-related example is hot hand in basketball. We notice certain pattern and ignore many equally probable others.

Probability Is Complex

Probability is a measure of "chances" or certainty about the occurrence of an event of interest. The measure could be objective or subjective or even a mixture of these. This means that there are more than one possible way for defining, choosing, or introducing such measures or for assigning probabilities. An important step in the study of probability was the introduction of rules or postulates that help with identification of inconsistent probability assignments. These rules provide guidelines leading to probability selections free of contradictions.

Suppose that to each event like A in the sample space S, a number "probability of occurrence of A," $P(A)$, is assigned, which obeys the following three postulates or axioms:

I. $P(A) \geq 0$

II. $P(S)=1$

III. $P(A \cup B)=P(A)+P(B)$, if A and B are incompatible (disjoint).

Using these, one can construct a deductive theory that will include all the rules needed to study a chance experiment.

Note that the above axioms do not completely determine the assignment of probabilities to outcomes. They only serve to rule out assignments inconsistent with our intuitive notions of probability. For example, when flipping a coin once, we assign probability to occurrence of a head. The probability of occurrence of a tail is one minus that number; no more, no less. So the question that arises at this point is, how do we assign probabilities to different outcomes of a chance experiment? In what follows, some frequently used definitions and methods are presented that could provide a partial answer to this question.

Probability Is Confusing: The Monty Hall Problem

Imagine a "Let's make a deal" type scenario. You are given the option to pick one of the three doors. Behind two of the doors is nothing, and behind the third door is a brand-new car. The goal of this game is, not surprisingly, to win the car. After you select your door, the host of the game will choose one of the other doors that has nothing behind it to open. After he does this, you are given the choice to keep your current door or switch to the remaining door. What do you do? But, more importantly, what should you do? What does probability tell you to do? The answer to this is somewhat counterintuitive. In fact, until computers were programmed to simulate this event, even world-class mathematicians refused to believe the result. The answer is that you should switch doors. At the beginning of the game, you have three choices and no information about any of them. The probability that you pick the car is 1/3, or 33 percent. You make your selection, and the host opens one of the other doors. Now two doors remain. It is tempting to think here that your chance of winning has increased to 50 percent, but this is not the case. Your probability remains unchanged. However, the probability of winning if you switch doors is 2/3, or 67 percent. If you do not believe me, suppose you switch. Then the only possible way you can lose is if you initially pick the door with the car behind it. The probability that you pick the door with the car from three choices is 1/3. So the probability that you lose if you keep your door is 67 percent, and since only one other door remains when you are offered the choice, the probability that you win if you switch is 67 percent.

Classical (Equally Likely or Theoretical) Definition

If n has only finitely many elements (outcomes), say n, and if by symmetry each of these elements (outcomes) has the same chance to occur–that is, if elements (outcomes) are equally likely– then an appropriate assignment for each possible outcome is $1/n$. If A has m elements (outcomes), then

$$P(A)=m/n = \text{Number of outcomes in } A/$$
$$\text{Number of possible outcomes.}$$

In practice, it is not always easy to establish whether simple events (outcomes) are equally likely or not. To resolve this, if we have no reason to say that simple events (outcomes) are not equally likely, we assume that they are. This is sometimes referred to as *principle of insufficient reason*. For example, if we assume a given die is a fair die (or have no reason to say that it is not), then the probability of event $A=\{2, 4, 6\}$ is equal to $3/6=1/2$.

It should be pointed out that the classical definition of probability has a logical problem in that it uses the probability (chance) in the definition of probability. Recall that equally likely means that outcomes have the same chance or probability of occurrence. Another problem relates to the fact that the only way to produce convincing evidence that a given coin is fair is by experimentation—that is, by flipping it. For fair coin, the relative frequency of the occurrence of event H is expected to tend to $P(H) = \frac{1}{2}$ if we perform the experiment (in this case flipping the coin) over and over (law of large numbers). Based on this, we may use the relative frequency as the probability of occurrence of an event *since* it is based on experiment. Calculation of probability based on this approach is discussed below.

Objective (Empirical) and Subjective (Judgmental) Definitions

Definition of probability based on relative frequency is usually referred to as "empirical probability or statistical probability." Formally, for an event A, it is defined as

Number of times event A has occurred / Total number of times experiment has been performed = Relative Frequency.

This definition of probability has wider applications compared to the classical definition.

This interpretation of probability is also referred to as objective interpretation because it rests on results of the experiments rather than any particular individual concerned with the experiment. In practice, this interpretation is not as objective as it might seem, since the limiting relative frequency (when an experiment is performed infinitely many times) of an event will not be known. Moreover, it refers to the experiments and events that are repeatable. Thus, we will have to assign probabilities based on our beliefs about the limiting relative frequencies. Note that there are many real-life situations where we have no control on occurrence of the event of interest. For example, we can create natural disasters such as an earthquake. In fact, in many cases, repetitions may not be possible. For instance, no frequency interpretation can be given to the event that New York Jets will win the Super Bowl next year. Other situations where this definition may not be applicable is when no or very few data regarding the event of interest is available. An example of the former is the probability calculation of a nuclear failure. For the latter case, if after six rolls of a die 4 is not yet observed, we cannot assign a probability $0/6 = 0$ to it since observing 4 has certainly a nonzero probability. For cases such as occurrence of an earthquake or failure of a nuclear power plant, we could only use expert's judgment and subjectively passing probabilities.

Summarizing these, probability is a number between 0 and 1 (inclusive) that provides an indication of how likely is an event to occur. Here, 0 refers to impossible and 1 refers to definite and everything else is, of course, between. Mathematics works mostly with items with probability 1. That is why you see theorems in mathematics books. Physics works with items with probability very close to 1. That is why you see laws in physics books. Then there are subjects such as biology, geology, etc., where you see theories or hypotheses, where probabilities are not very close to 1.

Which Definition of the Probability Should We Use?

So far, we discussed different definitions of probability. But when, for example, should we use equally likely definition? If we have no reason to doubt that outcomes are not equally likely, we will assume that they are. This, for example, applies to problems such as flipping a coin, rolling a die, drawing cards from a deck of cards, or choosing a student from a group of students randomly.

But it cannot be applied to, for example, weather prediction or being dead or alive tomorrow (we hope they are not equally likely), since, otherwise, we should expect half of the people dead by tomorrow. For these cases, relative frequency (statistics) is more appropriate and may be used. As a different example, think about buying insurance for your car. During a time period, you may make a claim or no claim. But can an insurance company assume that these are equally likely? Of course not. So insurance companies rely on statistics to find the probability that you will make a claim. The statistics they use includes things such as your age, type of the car, etc.

Now, if no past statistics are available or if an experiment is not repeatable (e.g., natural disasters), then one may consider a subjective assessment, such as an expert opinion, to assign probabilities to different possible outcomes. For example, what is the probability of complete failure of a nuclear power plant close to your home? For this, there is no past data (statistics), so experts may examine the components of the system and subjectively assign or estimate likelihoods. Then combine these likelihoods to arrive at a quantity presenting or estimating the risk (probability) of such events. In summary, subjective probabilities result from intuition, educated guesses, and estimates. Here, probabilities are considered to be measures of personal belief (or knowledge about the subject) and, hence, are usually different for two different people. As a sport example, given an athlete's extent of injuries, a doctor may think that he or she has 80 percent chance of full recovery. For the same athlete, a trainer with a different set of experiences may suggest a different probability estimate.

Finally, some words of caution about probability. First, probability is an undefined term, or it has many different meanings, to put it differently. The two most usual interpretations are (1) it is an intrinsic property of a physical system, and (2) it is a measure of belief in the truth of some statement. Second, there are many examples demonstrating the fact that assigning likelihoods to events, especially rare events, may lead to serious problems. Following one's intuition often results in numbers very different from the true likelihoods. For example, if you have thirty students in your class, what is the probability that at least two of them would have the same birthday? To anybody's surprise, more than 70 percent (why?). As a sport-related example, consider baseball and football hall of famers who share birthdays. Is this highly likely, or does it have a small chance? George Halas and Reel Schoedienst were both born February 2; Fran Tarkenton and Lang MacPhail, February 3; and Hank Aaron and Roger Stanback, February 5. Noting that these are just examples from early February, one can see that this is very likely event.

Table: Different Definitions of Probability

Type	Definition	Description
Classical (Theoretical)	Number of Outcomes in the Event Number of Possible Outcomes	Finite Number of Equally Likely Outcomes
Objective (Empirical)	Frequency of the Event Total Frequency = Relative Frequency	Based on Available Statistics or Relative Frequency From an Experiment
Subjective (Judgmental)	No Definition	Based on Intuition, Available Knowledge or Educated Guesses
Relationships: As the number of experiments is increased, empirical probability will approach the theoretical (actual) probability (the law of large numbers). As our knowledge about the matter under investigation is increased, our guess gets closer to the actual probability.		

Bayesian Approach

The Bayesian theorem is used in probability studies to help predict future events based on an event that already occurred, which is known as a prior probability. When data is not available, a prior distribution is used to calculate knowledge about the parameter. When data is available, we can update the prior knowledge using the conditional distribution of parameters. The Bayesian theorem makes it possible to transition from the prior to the posterior (Bayesian).

Some of the uses for the Bayesian theorem include medical diagnosing for predicting the outcome of the disease process and predicting how fast an infectious disease can spread. The theorem can also be used in sports to predict the outcome of a game and how many points a player might score during a game. Bayes' theorem can also be used to help predict odds in gambling. There are many other uses for Bayes' theorem. It can be applied whenever you need to predict future outcomes given previous data.

The Bayesians model is

$$P(A|B) = \frac{P(A \cap B)}{P(B)}.$$

By breaking down the model part by part, hopefully this will help you interpret why the Bayesian model helps predict future outcomes. I'm going to start breaking the model down by starting on the left side of the equal sign, which is $P(A|B)$. This part means the probability of A given (|) B—meaning, you want to find the probability of A given that you already have the information (data) for B. The second part is the numerator of the fraction $P(A \cap B)$. In this part, it makes it easier to set up your data in a table, tree diagram, or a Venn diagram so you can see what you need to multiply $P(A \cap B) = P(A) * P(B)$ (for independent events). The third part is the denominator of the fraction $P(B)$, which is the probability that you already know. Hopefully, this helps you understand why the Bayesian model helps predict outcomes, but if not, let use an example.

When betting that the second card drawn out of a deck will be a king given that the first card is a king, what will be your probability of winning being that the dealer is not replacing the cards as they are drawn?

$$A = \{2^{nd} \text{ card is a king}\} \qquad B = \{1^{st} \text{ card is a king}\}$$

$$P(A|B) = \frac{P(A \cap B)}{P(B)} = \frac{P(\text{1st card is a king and 2nd card is a king})}{P(\text{1st card is a king})} = \frac{\frac{4}{52} * \frac{3}{51}}{\frac{4}{52}} = \frac{3}{51} = 0.059$$

This means that you have about a 6 percent chance of winning.

Probability and Odds

Recall that if outcomes of an experiment are equally likely, then we may assign probability to an event A as

$$P(A) = \frac{\text{number of times A occured}}{\text{number of times trial is repeated}} \quad \text{or } P(A) = \frac{\text{number of outcomes in A}}{\text{total number of outcomes}}.$$

Odds

For such cases, odds are viewed as ratios of the number of outcomes in A and the number of outcomes in not A. We could also look at it as the number of successes to the number of failures, where a success is occurrence of event A. Formally,

$$\text{Odds } (A) = \text{Odds in favor of } A = \frac{p(A)}{1 - P(A)} = \frac{p(A)}{P(A)}.$$

For example, rolling a fair die, the odds of rolling a 3 are 1 to 5, or 1:5, as there is one way to roll a 3 (success), while there are five ways to not roll a 3 (failures). Note that since the probability of rolling a 3 is 1/6,

$$\text{Odds (Observing 3)} = \frac{P(3)}{1 - P(3)} = \frac{1/6}{5/6} = \frac{1}{5}.$$

Similarly, rolling two fair dice the possible outcomes are

$$(1,1), (1,2), (1,3), (1,4), (1,5), (1,6)$$
$$(2,1), (2,2), (2,3), (2,4), (2,5), (2,6)$$
$$(3,1), (3,2), (3,3), (3,4), (3,5), (3,6)$$
$$(4,1), (4,2), (4,3), (4,4), (4,5), (4,6)$$
$$(5,1), (5,2), (5,3), (5,4), (5,5), (5,6)$$
$$(6,1), (6,2), (6,3), (6,4), (6,5), (6,6)$$

Since there are three ways to roll sum 10, and thirty-three ways to not roll sum 10, the odds of rolling sum 10 is then 3:33. Note that that when the probability of success is low, the odds and the probability are very close.

A more interesting example is the ratio of odds for winning a home versus an away game.

$$\frac{P(\text{home win})/P(\text{home loss})}{P(\text{away win})/P(\text{away loss})},$$

which is expected to be larger than one. In fact, some recent investigations revealed that this ratio varies between 1.2 (for major league baseball) and almost 2.5 (for college basketball).

The odds of a lightning strike are 1:135,000. The odds of dying as a result of a meteor strike anywhere in the world, like the kind of rare but catastrophic geologic event that shapes an eon, are 1:75,000. The odds of winning the Powerball lottery are 1:195,249,054.

Even though the odds of dying from a meteor strike are better than winning the Powerball, many play the lottery and think that have a good chance of winning but hardly think that a meteor strike will affect them.

Odds and Gambling

Gambling has grown rapidly in the last few decades especially in relation to the sports competitions. People who gamble on outcomes of games and tournaments may be divided into three possibly overlapping categories:

1. The casual gamblers, who gamble for possible enjoyment may not be aware of odds or strategic subtleties and may not even be very knowledgeable about players and the teams.
2. The compulsive gamblers, who are happy primarily when in the act of gambling and do not care about the sport itself or even the outcome of the competitions.
3. The professional gamblers, who look to gambling as another profession. Such individuals understand odds, the games involved, and also know players, teams, and their potential well, although they may not have a good background in mathematics and the calculus of the odds.

One interesting fact is the following: many individual gamblers who understand the odds and their implications may not fully understand the difference between the true and house odds. To clarify this, consider the following table comparing the true and house odds for the game of roulette in a well-known place such as Las Vegas, Monte Carlo, and Atlantic City. For example, the true and house odds in Las Vegas roulette are given in table below.

Type of bet	True odds	House odds
Color (red or black)	20:18	1:1
Parity (even or odd)	20:18	1:1
18 #'s (1-18 or 19-36)	20:18	1:1
12 #'s (columns or dozens)	26:12	2:1
6#'s (any 2 rows)	32:6	5:1
4 #'s (any 4 numbers square)	34:4	8:1
3 #'s (any row)	35:3	11:1
2 #'s (adjacent)	36:2	17:1
Single #'s	37:1	35:1

If a $1 bet is made on red, then the chance of winning a dollar is 18/38; whereas, the chance of losing is 20/38. So, in the long run (or on average), one will lose $2 for every thirty-eight games he or she plays. With so many players and so many tables to play, it is not hard to imagine what could happen in the long run.

Turning to a sport example, betting works as follows: The bookmaker sets odds on each of the possible outcomes. Suppose that an odd of 39 to 1 is set for a player or a team. If this player wins, the person who has bet on this player winning will receive amount given in the odds, plus his or her money. This means a gain of $39. If the chance of winning for this player is 1/40, then the bet is called a fair bet. As in gambling clubs, if we convert the odds to the probabilities, we find that total probability assigned to the players or teams is less than one. The difference benefits the bookmakers and in long run makes them the winners. Remember that bookmakers have various ways to guarantee making money. People who know probability theory usually see no merit whatsoever in betting at racetracks, casinos, or in buying lottery tickets. After all, multimillion-dollar gambling places are built using gambling proceeds. Why would somebody bet with people who make a lot of money gambling?

Probability and Physics

The ideas of randomness are central too much of modern physics and have overthrown the "clockwork universe" conceptions of earlier centuries. The laws of probability and statistics were developed by such mathematicians as Fermat, Pascal, and Gauss, and received their first major application in physics in the kinetic theory of gases developed by Maxwell and Boltzmann.

Here, the use of probability is necessary because the number of particles involved is too great for a deterministic/mathematical calculation. With the advent of quantum theory, physics seemed to be based on an essential randomness, whose reality was debated by Bohr and Einstein until the end of their lives. Only later, in the experiments of Alain Aspect, has a convincing demonstration been given that the inescapable randomness of quantum theory is a fact of nature.

Since the molecules and their collisions are so numerous, and the velocities so varied, and since our ignorance of the initial conditions is almost total, Maxwell postulated that positions and velocities are distributed at random; and he was confident that this assumption would describe the gas adequately and would allow one to calculate the mean values of the macroscopic variables. His breathtaking intuition was confirmed half a century later by the work of Albert Einstein (1905) and of Jean Perrin (1908) on Brownian motion.

To clarify, think about one mole of gas. One mole of any molecular substance such as $O2$ contains $6.02 \times 1,023$ molecules. To construct the theory governing a deterministic system of $1,023$ molecules, physicists exploited probabilities through ignorance, and with complete success. The reason for this success deserves discussion. The Soviet physicist Lev Landau has shown that a classical system requiring infinitely many parameters would behave in a totally random fashion; in other words, it would be random unavoidably, and not merely by reason of our ignorance.

Boltzmann bases his analysis on the following observation: the molecules are so fast and their collisions so frequent that the system rapidly loses or at least appears to lose track of the initial conditions. Typically, this leads us into the realm of probabilities through ignorance. His based his theory on the following postulates:

1. Every molecule has equal a priori probability of being in region A.
2. The system evolves spontaneously from the less toward the most probable state.

This type of studies led to model for phenomenon such as Brownian motion. The phenomenon was discovered and studied first by the botanist Robert Brown in connection with the erratic motion of pollen grain suspended in fluids. Einstein first presented a quantitative theory of the Brownian motion in 1905 based on kinematic theory and statistical mechanics. He showed that the motion could be explained by assuming that the immersed particle was continually being subjected to bombardment by the molecules of the surrounding medium. Wiener developed rigorous mathematical explanation in 1918 based on stochastic process called Wiener-Levy Process.

Examples

- A dust particle suspended in water moves around randomly, executing what is called Brownian motion. This stems from molecular agitation, through the impacts of water molecules on the dust particle. Every molecule is a direct or indirect cause of the motion, and we can say that the Brownian motion of the dust particle is governed by very many variables. In such cases, one speaks of a random process; to treat it mathematically, we use the calculus of probabilities described.

- A compass needle acted on simultaneously by a fixed and by a rotating field constitutes a very simple physical system depending on only three variables. However, we shall see that one can choose experimental conditions under which the motion of the magnetized needle is so unsystematic that prediction seems totally impossible. In such very simple cases whose evolution is nevertheless unpredictable, one speaks of chaos and of chaotic processes; these are the terms we use whenever the variables characterizing the system are few.

Deterministic method allow one to predict the future exactly from initial conditions that are likewise exact. The best example of a deterministic theory is classical mechanics. However, this definition of determinism is based on the tacit assumption that the deviations on arrival diminish roughly in proportion to the deviations at departure. In that case, the idealized limit can be envisaged quite clearly; and in fact, gunners can realize excellent approximations to it. But what would happen if initial and final deviations were connected by a relation more complex than simple proportionality?

Misleading Use of Probability

Mark Twain, in his book published in 1924, mentions a famous line attributed to Benjamin Disraeli, "There are three kinds of lies; lies, damned lies, and statistics." Also, we frequently hear the sayings that "that is just statistics" and that "you can prove what you wish to prove with statistics." In fact, both statistics and statisticians have a poor image in the minds of many people. This is because most data can easily be manipulated in an unethical and unscientific fashion to draw desired conclusions. In other words, it is easy to distort the truth. This is because most people are unfamiliar with concepts and the language of statistics.

An interesting comparison with medical science and doctors is as follows. If you do not find your doctor's instructions helpful, most probably you will blame him or her, not the medical science. But when it comes to statistics, most people blame the science or the methodology used, not the user. There are good and bad engineers, and good and bad lawyers, despite the fact that both professions require a license to practice.

One problem that arises when using probability is that most of us in our daily decisions rely on our intuition about probability–the intuition that has been cultivated over many years, based on a wide range of experiences. Unfortunately, our intuition can be completely unreliable. For example, why do people tend to think that parents with four baby girls in a row are due to have a baby boy? Why do we often get more concerned about low risks like nuclear power plants than high risks like driving? Why do people tend to think a lottery ticket with the numbers 17, 15, 18, 19, 44, 51 has a better chance of winning than one that has the numbers 9, 8, 7, 6, 5, 4 when both actually have the same chance? Researchers have studied questions like these for many years, and their discoveries are illuminating.

Fortunately, in sports and the related contexts, people rely on statistics more than anything else. But as is discussed in several sections of this book, the way statistics are presented may be misleading. A good example of this is Simpson's paradox discussed in section 2.7 related to what is called hot hand. Here, we may illustrate the point by presenting a different example.

When writing this section. it was middle of December 2001. Looking at NBA records, I noticed that almost all the teams in the eastern division had a double-digit number of wins and losses. So it was hard to pick a team as the best. I thought this can be related to the discussion of this section as follows.

Suppose that in the area you live, people like to bet on outcomes of the games played between the teams in a particular tournament or conference (in this case, eastern division). You can take, for example, sixty-four of these people and send them a letter with a forecast for the next six games. Since there are a total of sixty-four possible outcomes such as WLLWWL, LLWWWL, etc., you can send each of these outcomes to one individual. By doing so, one person will get all the forecasts correct, and six people will get five correct forecasts. These people may then start believing that you must have some special information about the games. After all, the probability of correctly guessing outcomes of six successive tosses of a fair coin is only.5x.5x.5x.5x.5x.5=.0156. If each week you do this by sending forecasts to several different groups of sixty-four people and start new groups each week, you may be able to create reputation and generate clients to make significant profit.

One real story regarding the prediction is related to the price of an average family home in the year 2000. In mid-1980s, housing prices in a large part of United States rose by more than 20 percent per year. Based on this, some "experts" suggested that the price of an average family home would exceed $1 million. However, when year 2000 arrived, prices were nowhere near their prediction. This, again, led some people to doubt and blame statistical forecasting, not the so-called experts. A similar thing happened to stock market where experts were predicting 15,000 for Dow Jones Industrial Average and 6,000 for Nasdaq for the year 2002, while, in fact, the opposite happened.

Summarizing these, statistics do not lie, but people do. This science together with the methodology used, like a paring knife, can be a very valuable tool when it is properly used. When it is misused, it can lead to some very bad results. Some of the ways to intentionally or inadvertently distort the truth with statistics are described in the book *How to Lie with Statistics*, by D. Huff (Norton 1954, New York).

Paradoxical Probabilities

Examples mentioned in this section were presented earlier. However, as they are relevant to the discussion of this chapter, they are repeated here.

To reason correctly, when we use probabilities, it is difficult and requires thinking that seems contrary to intuition. For example, consider the following hypothetical situation concerning a diagnostic test for cancer.

Suppose that in the over-thirty-year-old population in the United States, 0.5 percent, one out of two hundred people, actually have cancer. Suppose further there is a test for cancer that is 98 percent accurate: if the test is performed on persons who have cancer, the test results will be positive 98 percent of the time, and if the test is performed on persons who don't have it, the test results will be negative 98 percent of the time. Table 4 shows the expected results if this test is performed on ten thousand people (of whom about fifty actually will have cancer).

Hypothetical results of 98 percent accurate cancer detection test

	Number of positive test results	Number of negative test results
People free of cancer (9950)	199	9751
People having cancer (50)	49	1
TOTAL	248	9752

Suppose you have had the test and the results come back positive. Do you have cancer? Probably you would reason that since the test is 98 percent accurate that it is almost certain that you do.

However, a positive test result is wrong for most people and you probably would not have cancer. Look back at table 4. What portion of the persons with positive test results actually have cancer? The answer is 49/248, or slightly less than 20 percent. Although the test is 98 percent accurate, since most people are healthy, most of the test's errors are on healthy people, and a positive test is likely to be in error.

Here is a second counterintuitive probability example. This one deals with the unlikely event that two randomly chosen people will share the same birthday. Surprisingly, the likelihood of coincident birthdays becomes high in groups that are much smaller than 365.

A particular professor with an announced policy of occasional surprise quizzes comes in to meet her class of thirty students and asks each student to write down his or her birthday month and day, ignoring the year, on a slip of paper. She collects the slips and then announces that there will be a surprise quiz today if two of the slips name the same birthday. As preparations are being made to tally the dates, do you suspect that the students are already beginning to relax, confident that with only thirty students and 365 days to choose from, there will not be a match?

Although we cannot say for certain what will happen in this particular class, elementary concepts of probability theory and a straightforward calculation can be used to discover that the probability of a repeated birthday in a group of thirty people (with no twins and no February 29 birthdays) exceeds 70 percent. In fact, the probability of a matching birthday is over 50 percent for a group of people as small as twenty-three.

Risk Presentation

Television, reflecting great public interest and concern, provides extensive coverage of natural disasters with strong visual impact. Yet television is often criticized for the way it presents risks. To date, very few attempts have been undertaken to assess the role of the broadcast media in elaborating and reinforcing dominant public perceptions of risks during disasters.

Flying versus Driving

Some people think that traveling by plane is inherently more dangerous than driving a car. According to the National Safety

Council, during the life of a randomly selected person, the odds of dying in a motor vehicle accident and in air and space transport are 1 in 98 and 1 in 7,178, respectively. This indicates that flying is far safer than driving. However, for some, flying may feel more dangerous because our perception about risk is usually formed based on factors beyond mere facts. For example, most people think that they are good drivers and, as such, feel safer because driving affords personal control. Plane crashes are often catastrophic. It kills many at once and grabs the attention of major media, which make people more sensitive to them.

Are You an Average Person?

In general, there is a lot more to calculating and comparing risk than one might think. According to the experts, for an average American, the annual risk of being killed in a plane crash and a motor vehicle are about one in eleven million and about one in five thousand, respectively. But is this all? First, most of us are not the average American. Some people fly more than others, and some do not fly at all. So if we take the total number of people killed in commercial plane crashes and divide that into the total population, the result gives a good general guide, but it is not specific to our personal risk. Here are other useful numbers:

1. Dividing the number of people who die into the total number of people gives us the risk per person.
2. Dividing the number of victims into the total number of flights passengers took gives the risk per flight.
3. Dividing the number of victims into the total number of miles all of them flew gives you the risk per mile.

1995 Data

In 1995, out of every 100 million, about 16,300 and 111 people were killed in automobile and commercial flights accidents, respectively. The number of deaths per one hundred miles were, respectively, 3 and 100 for 100 million miles traveled and 30 and 20 for every 100 million trips made.

This shows that the risk of death per mile is 33 times higher for car. However, the risk of death per trip is about 1.5 times higher for airplanes.

Discussion

All the above calculations produce useful and accurate numbers. However, which one is most relevant to us depends on our personal flying patterns. Some fliers take many short flights. Some take less but longer flights, for example. Since the overwhelming majority of the plane crashes take place in connection with takeoffs and landings, the risk is less a matter of how far you fly and more a matter of how often. If you are a frequent flier, then the risk per flight means more to you. For occasional long-distance fliers, the risk per mile means more. A frequent long-distance flier would want to consider both.

The number of plane crash fatalities in the United States varies widely from year to year. So calculation of risk based on one year and average of five years, or ten or twenty, may be different significantly. In some years, no plane crashes occur, or at least very few do. This makes the value of the risk per year misleading. If we average things over, say, five years, or ten, some other factors will muddy the waters. In the last five years, safety factors have changed. A ten-year average might be misleading too.

Despite all these caveats, numbers are a great way to put risk in general perspective. Without question, most metrics show that flying is less risky than traveling by cars. But wait: Just when you thought it was safe to use numbers to put risk in perspective . . . Numbers are not the only way—not even the most important way—we judge what to be afraid of. Risk perception is not just a matter of the facts. For example, consider the risk awareness factor. The more aware of a risk we are, the more concerned about it we become. This explains why, when there is a plane crash in the news, flying seems scarier to many of us, even though that one crash has not changed the overall statistical risk significantly.

CHAPTER 10

Rare Events

**Rare events often occur more than they
should, and that is the problem.**

Averages versus Extremes

TRADITIONALLY,
STATISTICS EDUCATION
has focused on the study of values with high frequencies and on
averages. This is evident from examination of course descriptions
and contents of textbooks on introductory statistics. As was pointed
out earlier, this is because the general approach to statistical
modeling has been to treat data as a message and decompose
that into a systematic part and a random part. Averaging, which
isolates the smooth part, is an integral part of this approach. In
many applications, however, it is impractical to focus on averages
alone. In fact, it has become important to study the extreme
values and rare events since these values are usually accompanied

by severe consequences. Clearly, large earthquakes are of more concern than average earthquakes, and this is true for other natural disasters. In examination of strength of materials and of system reliability, it is the weakest links that leads to failure. In the study of a parallel or series systems, components with maximum or minimum resistance or lifetime are of greatest concern. In risk management, we must be prepared for the largest disasters; our insurance must be designed to cover the largest claims. Finally, in sports, we are interested in records—that is, the largest values or shortest times. Recall that extreme events and rare events are often newsmakers and capture the attention of the public.

The celebrated center limit theorem has given statistics its focus on averages for we do what we know how to do. The theories of extremes and record values are less simple, less unified, and more recent but not less important. In light of this, we think that there is a need for inclusion of extreme values and related theories in introductory probability and statistics courses, despite their difficulties.

Rare Events

This part aims to explore the theory regarding the low-probability events such as exceedances and extreme values with a particular focus on its application. It deviates from material taught in introductory statistics courses in that it concentrates on frequency and values of extremes and the situations where such values are of greater concern than averages. The theory of exceedances, together with a brief account of extreme value theory, threshold theory, and theory of records, is presented. It is hoped the instructors of statistics will find this part of some value for teaching statistics and for its focus on a popular motivating theme.

Traditionally, statistics has focused on the study of averages and typical values with high frequencies. This is evident from examination of course descriptions and textbooks for introductory statistics courses offered in universities and colleges. Most classical approaches treat the data as a message and seek to decompose it into a systematic part (signal or trend) and a random part (noise). The techniques such as smoothing are then used to recover the systematic part. Smoothing usually is achieved by some sort of averaging (low-pass filtering). The difference between the original message and the systematic part (i.e., the random part) is then used to provide a reliability statement regarding the systematic part.

In wider applications, however, it is becoming increasingly impractical to focus on averages alone. It has become important to understand the distributions of extreme values and rare events since such events are usually accompanied by severe consequences. Consider, for example, a natural disaster such as an earthquake. Clearly, a moderate earthquake is of less concern than a severe one. In the reliability study of systems and structures, it is the weakest link that is worrisome. In risk management and insurance industry, the highly complex situation and possible large claims are what one needs to prepare for. The celebrated *central limit theorem* has given statistics its focus on averages–for we do what we know how to do. The theories of extremes and record values are less simple, less unified, and more recent–but not less important.

Extreme values are usually analyzed using one of the three major theories. These theories are concerned with the actual values of the extremes. The following is a brief description of each:

1. The *extreme value theory* usually deals with annual maxima or minima. This theory is limited to the absolute largest or absolute smallest data value in a specific period (e.g., a year).

2. The *threshold theory* deals with values above or below a specified threshold.

3. The *theory of records* deals with values larger or smaller than all previous values.

As pointed out, these theories deal with the values of the extremes. The frequencies of extremes are usually analyzed using a theory known as the *theory of exceedances*. This theory deals with number of times a chosen threshold is exceeded.

In most applications, the number of exceedances and values of excess over a threshold are combined to yield a more detailed analysis. For example, insurance companies analyze both the number of times a "large" claim is made and the amount by which the claims exceed a large threshold. Reliability engineers study both the number of large earthquakes and their sizes. We apply the same methodology to sports.

Exceedances and Excesses

In many applications such as sports, extraordinary individuals or teams are those who pass a threshold classified as average and ordinary. For example, in one-hundred-meter dash for men, only athletes who run the distance in less than ten seconds may be considered exceptional. In basketball, a player with a point per minute or assist per minute greater than 2/3 may be referred to as exceptional. In tennis, players who win several grand slam or have more than forty-five aces in one match may be considered exceptional. If we examine the distribution of times for one-hundred-meter race for the runners or points per minute (PPM) for basketball players, then we find that the exceptional performances constitute the tail of the corresponding distribution (for one hundred meters times, the lower tail; and for PPM, the upper tail). Thus, to study such performances, rather than entire distribution covering the entire range of possible values, we may just study the upper (lower) tail of the corresponding distribution. We can do this either by seeking a model for values above (below) a chosen threshold, or by considering, for example, the third-best performance in the history of that sport as threshold and study the number of times it was surpassed or exceeded. This type of analysis allows us to find how many basketball players would it take to produce a player that will perform like or better than, for example, Michael Jordan.

Formal Definition of Exceedances

Assuming independent and identically distributed (*iid*) trials, determine the probability of r exceedances in the next n trials. For example, in a season of n games, how many times will a basketball player or a team score more than x points?

This situation can be described by a Bernoulli experiment with two possible outcomes: "exceedance" or "not exceedance." Since the experiment is repeated n times, the number of exceedances has a binomial distribution with parameters n and $p(x)$ where $p(x)$ is the probability of exceedance of the level x of the variable of interest. Note that $p(x)=P(X>x)=1-F(x)$, where $F(x)$ is the cumulative distribution function of X.

How Often? Return Periods

Consider an event such as breaking a record whose probability of occurrence in a unit period (normally one year) is p. Assume that occurrences of such events in different periods are independent. Then, as time passes, we have a sequence of Bernoulli experiments with two possible outcomes: *(a)* occurrence or *(b)* nonoccurrence. Thus, the time (measured in units of a period of interest) to the first occurrence is a geometric random variable with parameter p and a mean value of $1/p$. This motivates the following definition.

Definition. Let E be an event, and T the random time between consecutive occurrences of E. The mean value of the random variable T is called the return period of the event E.

Note that if $F(x)$ is the distribution function of the yearly maximum of a random variable, the return period of that random variable to exceed the value x is $1/[1-F(x)]$ years. The return period of the variable to go below the value x is $1/F(x)$ years.

A Point to Consider

The above results are applicable when observations form a sequence of independent and identically distributed random variables. This assumption is not reasonable for most cases in which records are broken more or less frequently than what theory predicts. One way to deal with this problem is to assume that every year more/less events take place, resulting in a higher chance of setting a record.

To clarify, suppose that an event starts with one hundred participants and the rate of participant growth is 4 percent. This means that with each passing year, more participants would enter, thereby increasing the chance of setting a new record. The number will increase to $100(1.04)^{10} = 148$ after ten years and to $100(1.04)^{20} = 219$ after twenty years. If we assume that each participant has an equal chance of breaking a record, the chances of breaking the original record after ten and twenty years are 1.48 and 2.19 times of that for starting year, respectively. Note that using the past history, we can estimate both the number of first-year participants and the geometric rate of growth. Alternatively, we can choose an $a(t)$ such that as time passes, the chance of setting a record increases.

Discussion

This part aims to emphasize the need for inclusion of statistical theories of extreme values in introductory statistics courses. It points out situations where extremes values are of greater concern or importance than averages. The theory of exceedances together with a brief account of extreme value theory, threshold theory, and the theory of records are discussed. It is hoped the instructors of statistics find this part of some value for teaching topics related to non-typical and non-average values.

Traditionally, statistics has focused on the study of values with high frequencies and on averages. This is evident from examination of course descriptions and textbooks for introductory statistics courses offered in universities and colleges. In the modern world, however, it is becoming increasingly impractical to focus on averages alone; it has become important to pay attention to rare events and extremes, since they are newsmakers and are often accompanied by severe consequences.

The celebrated *central limit theorem* has given statistics its focus on averages–for we do what we know how to do. The statistical theories of extreme values are less simple, less unified, and more recent. However, they are not less important. In fact, there is a need for inclusion of extreme value statistics in introductory courses.

A Summary

Why Averages?

As pointed out, most textbooks focus on values with high frequencies and on averages. This is because most classical approaches treat the data as a message and seek to decompose it into a systematic part and a random part. Another representation is

Message = Signal (systematic part) + Noise (random part).

For data with time index, known as times series, the Wald decomposition theorem states the following:

Time Series = Deterministic (Trend) + Stochastic (Noise).

When the form of a signal is known or is assumed, it is easy to separate the two parts. In the absence of information concerning the systematic or random parts, smoothing is used to recover each part. Smoothing is an exploratory operation, a means of gaining insight into the nature of data without precisely formulated models or hypotheses. This is usually carried out assuming that the systematic part is smooth and the random part is rough—that is,

Message = Smooth part (trend) + Rough part (noise).

Smoothing is often achieved by some sort of averaging (low-pass filtering). Several techniques (low-pass filters) are available for doing this. Once the smooth part is determined, the difference between the original message and the smooth part is used to provide an estimate for the rough part. The rough part is usually used to make reliability statement regarding the systematic part.

For time series, for example, one popular smoothing technique is known as moving averages. The idea is to average the neighboring values and to move it. To clarify, suppose that $Y_t, t=1, 2,...,n$ is an observed time series of length n. The moving average of order $2p+1$ is defined as where $w_j > 0$ are weights such that

$$\sum_{j=-p}^{p} wj = 1.$$

The weights are usually selected based on prior information. One classical approach, referred to as exponential smoothing, chooses the weights

$$S_t = \sum_{j=-p}^{p} w_j Y_{y-j}, t = p+1,...,n-p$$

according to a geometric progression. It is useful when assigning bigger weight to more recent values make sense.

Why Extreme Values?

In many applications, it is not appropriate to focus on averages alone. In fact, there are many instances where extreme values and values with low frequencies are of more concern than average values or values with high frequencies. Examples include a large natural disaster compared to a moderate or an average one, a weak component or link compared to their average counterpart, or a large insurance claim compared to an average claim. In sports, of course, fans only remember out-of-ordinary or exceptional performances or games.

So one may ask, why then extremes have not yet found their place in classical textbooks? There are several reasons for this: maybe because the celebrated *central limit theorem* has given statistics its focus on averages; or because the theories for extremes are less familiar, less unified, and more complex. Moreover, they are more recent, and some aspects of these theories are not yet fully developed.

As pointed out earlier, extreme values are usually analyzed using one of the three major theories:

1. the extreme value theory, which deals with "annual" maxima or minima
2. the threshold theory, which deals with values above or below a specified threshold
3. the theory of records, which deals with values larger or smaller than all the previous values.

These theories deal with the actual values of the extremes. The frequencies of extremes are analyzed using the theory of exceedances. This theory deals exclusively with number of times a chosen threshold is exceeded. For example, in track and field, athletes may be required to meet or exceed a pre-specified performance levels to qualify for major competitions. Here, we may be concerned with the number of the athletes that succeed. Alternatively, we may be interested in a level that ensures a certain number of qualifiers. Note that in most applications, the number of exceedances and values of excesses over a threshold are combined to yield a more detailed analysis. For example, insurance companies analyze both the number of times a "large" claim is made and the amount by which these claims exceed a specific large threshold. Reliability engineers study both the number of extreme loads, such as large earthquakes, and their magnitudes. In what follows, a brief account of each theory is given.

Extreme Value Distributions

Extreme value theory generally deals with the annual (or any other period) maxima or minima. Specifically, the theory is based on dividing the sample into a subsamples and fitting a distribution to maxima or minima of the subsamples. For example, data may consist of largest earthquakes in California for each of the last one hundred years.

Here, like most statistical theories, first, distribution of the largest or the smallest values were derived for a finite sample. Then by letting sample size tend to infinity, the limiting distribution of extreme values were obtained. For this, a typical maxima Y_n is reduced with a location parameter β_n and a scale parameter α_n (assumed to be positive) such that the distribution of standardized extremes $(Y_n - \beta_n)/\alpha_n$ is nondegenerate. The forms of the limiting distributions are specified by the extreme value theorem. This theorem states that there are three possible types of limiting distributions (denoted by $F_y(y)$) for maxima:

1. The Gumbel distribution (type I) for which
   $$F_Y(y) = \exp(-\exp(-y)) \text{ for } -\infty < y < \infty$$

2. The Fréchet distribution (type II) for which
   $$F_Y(y) = \begin{cases} 0 & \text{for } y \leq 0 \\ \exp(-y^{-k}) \text{ for } y > 0 \ (k>0) \end{cases}$$

3. The Weibull distribution (Type III) for which
   $$F_Y(y) = \begin{cases} 0 & \text{for } y \leq 0 \\ \exp(-y^{-k}) \text{ for } y > 0 \ (k>0) \end{cases}$$

These three forms can be combined to yield the generalized extreme value distribution taking the form

$$P(\max(Y_1, Y_2, \ldots, Y_N) \le y) = \exp\{-\lambda(1-ky/\sigma)^{1/k}\},$$

Where, depending on whether parameter k is positive, zero, or negative, we get type I, type II, or type III extreme value distribution, respectively.

Most classical distributions fall in the domain of attraction of one of these three types. For example, distribution of maxima of samples from a normal distribution tends to Gumbel distribution. The necessary and sufficient conditions for a particular distribution to fall in domain of attraction of one of the three types. For type I, it is

$$\lim_{t \to +\infty} n[1 - F_Y(\alpha_n t + \beta_n)] = e^{-y}.$$

For type II, it is

$$\lim_{t \to +\infty} \frac{1 - f_Y(t\ y)}{1 - f_Y(t)} = y^k,$$

where $t > 0$ and $k > 0$ and $f(.)$ denotes the density function. Finally, for type III, it is

$$\lim_{t \to 0} \frac{1 - F_Y(t\ y + u)}{1 - F_Y(t + u)} = y^{-k},$$

where u is the endpoint of the distribution for Y ($F_Y(u) = 1$), $t > 0$, and $k < 0$.

Basic results obtained are the following:

- Only distributions unbounded to the right can have a Fréchet distribution as a limit.
- Only distributions with finite right end- point ($u < \infty$) can have a Weibull as a limit.

- The Gumbel distribution can be the limit of bounded or unbounded distributions.

How do we decide which of the three limiting distributions fit the data? Theoretically, we can use the fact that each of the classical distributions falls in the "domain of attraction" of one of the limiting distributions above. This works if distribution of the original data is known. Unfortunately, in practice, such information is not usually available, and decisions should be based on the area of application or on expert opinion. For example, in the case of sports, one may think of a possible ultimate record; in which case, the distribution bounded above (below) is more appropriate. When information about the appropriate limiting distribution is absent, statistical goodness of fit may be used.

When fitting extreme value distributions to annual maxima or minima, it is possible to discard some relevant data related to the years with several large observed values and retain less informative data from the years with no real large values. The threshold theory discussed next avoids this problem.

Summary

Assessing the probability of rare and extreme events is an important issue in the risk management of financial portfolios. Extreme value theory provides the solid fundamentals needed for the statistical modelling of such events and the computation of extreme risk measures.

Extreme value theory (EVT) is a branch of statistics dealing with the extreme deviations from the median of probability distributions. There exists a well-elaborated statistical theory for extreme values. It applies to (almost) all (univariate) extremal problems. From EVT, extremes from a very large domain of stochastic processes follow one of the three distribution types: Gumbel, Fréchet/Pareto, or Weibull.

The generalized extreme value (GEV) distribution is a family of continuous probability distributions developed within EVT. The GEV combines three distributions into a single framework. The distributions are

Type I: Gumbel   Type II: Fréchet   Type III: Weibull

The GEV allows for a continuous range of possible shapes. The shape parameter, S, governs the tail behavior of the distribution. The subfamilies defined by $S \sim 0$, $S > 0$ and $S < 0$ correspond, respectively, to the Gumbel, Fréchet, and Weibull families. Note the differences in the ranges of interest for the three extreme value distributions: Gumbel is unlimited, Fréchet has a lower limit, while the reversed Weibull has an upper limit.

The GEV facilitates making decisions on which distribution is appropriate. The GEV distribution is often used as an approximation to model the minima or maxima of long (finite) sequences of random variables. In general, the GEV distribution provides better fit than the individual Gumbel, Fréchet, and Weibull models. For example, in most hydrological applications, the distribution fitting is via the GEV as this avoids imposing the assumption that the distribution does not have a lower bound (as required by the Fréchet distribution).

Generalized Pareto Distribution

The threshold theory allows one to make inference about the values above or below a threshold—that is, the upper or the lower tails of a distribution. It considers the excesses, the differences between the observations over the threshold, and the threshold itself. Like extreme value distributions, there are three models for tails:

1. Long-tail Pareto
2. Medium-tail exponential

3. Short-tail distribution with an endpoint

Again, most classical distributions fall in domain of attraction of one these tail models. It has been shown that the natural parametric family of distributions to consider for excesses is the generalized Pareto distribution (GPD)

$$H(y;\sigma,k)=1-1(1-\frac{ky}{\sigma})^{1/k}.$$

Here, $\sigma > 0$, $-\infty < k < \infty$ and the range of y is $0 < y < \infty$ $(k \le 0)$, $0 < y < \sigma/k$ $(k > 0)$.

This is motivated by the following considerations:

- The GPD arises as a class of limit distributions for the excess over a threshold, as the threshold is increased toward the right-hand end of the distribution (i.e., the tail).

- If Y has the distribution $H(y, \sigma)$ and $y' > 0$, $\sigma - ky' > 0$, then the conditional distribution function of $Y - y'$ given $Y > y'$ is $H(y; \sigma - ky', k)$. This is a "threshold stability" property; if the threshold is increased by an arbitrary amount y', then the GPD form of the distribution is unchanged.

- If N is a Poisson random variable with mean λ and Y_1, Y_2, ..., Y_N are independent excesses with distribution function $H(y; \sigma, k)$, then
$$P(max(Y_1,Y_2,\ldots,Y_N) \le y) = \exp\{-\lambda(1-ky/\sigma)^{1/k}\},$$

which is the generalized extreme value distribution. Thus, if N denotes the number of excesses in, say, a year and Y_1, Y_2,..., Y_N denote the excesses, then the annual maximum has one of the classical extreme value distributions.

- The limit k → 0 of the GPD is the exponential distribution.

In practice, the proposed method is to treat the excesses as independent random variables and to fit the GPD to them. The choice of threshold is, to a large extent, a matter of judgment depending on what is considered large or small.

The theory is very useful when modeling large values based on observed large values is of main concern. Clearly, the modeling and prediction of large earthquakes should be based on past large earthquakes, not on past medium or small earthquakes. The same is true for sports records and performances as moderate values do not carry information about the exceptional future performances.

Summary

The relation between records and their frequencies for sports events has been considered by many investigators. Here, like many other areas of statistical application, we are primarily interested in drawing inference about the extreme values of a population of records or performances. The most commonly used method, developed at length by Gumbel, is to divide the data into subsamples and to fit one of the extreme data points in each subsample. In many cases such as environmental series, the natural subsample is one year's data; in fact, Gumbel's method is often referred to as the method of annual maxima. This also makes sense for sports since here, too, usually, annual or seasonal records are of interest. However, in other areas of application such as earthquake engineering, there is no natural seasonality in the data, and the subsample method appears artificial and wasteful. This may also be the case for some sports because of the expansion of sports-related activities and the fact that seasons do not happen at the same time all over the world. An example of this is tennis for which there are tournaments all year round. When applying to a given sport, subsample method presents a further difficulty. To clarify this, consider a sport such as men's two-hundred-meter race and take the period to be one year. Gumbel's method fits one of the extreme value distributions to the data presenting the best time for each year. Now, suppose that in a given year (an exceptional year), more than one record was set. Gumbel's method drops these records and uses only the best for that year together with the best records for other years, which could not be as good as the second-best record of the year in which more than one record was set. This means that a great deal of relevant information regarding the records (extremes) may be ignored by this method. With this and other drawbacks of the methods used for prediction of the extreme values, consideration of an alternative approach that avoids some or all of these difficulties is much needed. Here, we consider a theory concerning the tail behavior of statistical distribution introduced by Pickands (1975). This approach is based on fitting a suitable parametric model to a few of the largest order statistics corresponding to the best records regardless of when they were set. Since this approach has several advantages over the traditional subsample method, in recent years, a great deal of effort has been put in by various investigators to develop the theory and the methods based on that to its fullest potential. These have led to introduction of new classes of alternative procedures commonly referred to as "the threshold method." In fact, so far, two main procedures have been made available for practical applications: one by hydrologists called POT (peaks over threshold) method, and one by statisticians based on the GPD (generalized Pareto distribution) and the use of extreme order statistics. These two areas of work may be regarded as contributions to the same general problem, —that is, the modeling of the extreme characteristic of a series of observations in terms of its exceedances (in fact, excesses) over a high (low) threshold level.

Because of the importance of the results concerning the tail distribution and also power and flexibility of the methods based on them, this chapter is devoted to threshold method and discussion of its application to sport data.

References

Davis, R. and S. Resnick. 1984. "Tail Estimates Motivated by Extreme-Value Theory." *Annals of Statistics* 12: 1467-1487.

De Haan, L. 1981, "Estimation of the Minimum of a Function Using Order Statistics." *Journal of the American Statistical Association* 76: 467-469.

Hill, B. M. 1975. "A Simple General Approach to Inference about the Tail of a Distribution." *Annals of Statistics* 3: 1163-1174.

Pickands, J. 1975. "Statistical Inference Using Extreme Order Statistics." *Annals of Statistics* 3: 131-199.

Smith, R. L. 1988. "Forecasting Records by Maximum Likelihood." *J. of the Amer. Stat. Assoc.* 83: 331-338.

Records

Theory of records deals with values that are strictly greater than or less than all previous values. Suppose that data consists of the real numbers $Y_1, Y_2, ... Y_n$ with Y_n representing the most recent measurement. Usually, Y_1 is counted a record, as it is the largest value at the starting point. Y_i is a record (upper record or record high) if it is bigger than all previous values or measurements–that is, if

$$Y_i > \max(Y_1, ..., Y_{i-1}) \text{ for } i \geq 2.$$

The study of such values, their frequency, times of their occurrence, their distances from each other, etc., constitutes the theory of records. Formally, the theory of records deals with four main random variables:

1. The number of records in a sequence of n observations
2. The record times
3. The waiting time between the records
4. The record values

It is interesting to note that the first three can be investigated using nonparametric methods; whereas, the last one requires parametric methods.

Records in general and sports records in particular are of great interest, and their occurrence usually results in a great deal of media attention. Examples include the chase of the single season home run record by baseball players and breaking of the men's one-hundred-meter record.

The mathematics behind the theory of records is both interesting and elegant. Think about total annual snowfall in the city you live in the next ninety or one hundred years.

Q: What is the chance that a baby born this year would experience, for example, six personal record snowfalls? ten personal record snowfalls?

Q: How many personal record snowfalls should she expect to experience?

Q: How many record snowfalls is the most likely case?

The first year is a record. No history. What about the second (next) year?

The chance that the second year is a record is 50 percent, or 1/2. What about the third year?

The chance that the third year is a record is 1/3.

R_n: Number of records in n years

Term Years	Expected Number of Records
1	1
2	1/2
3	1/3
4	1/4

$$n \qquad \frac{1/n}{E(R_n) = 1 + \frac{1}{2} + \frac{1}{3} + \frac{1}{4} + \ldots + \frac{1}{n}}$$

For $n=90$

$$E(R_n) = 1 + \frac{1}{2} + \frac{1}{3} + \frac{1}{4} + \ldots + \frac{1}{90} = 5.08 \approx 5.$$

Note: 5 is also the most likely scenario (probability $= 0.21$).
Every record will eventually be beaten.

In fact, surprisingly, the intuitive idea that every record will be beaten also leads to mathematical proof that the harmonic sum

$$1+1/2+1/3+\ldots$$

grows without bound, becoming bigger than any finite number.

An Application

Values of n such that $E(R_n) \geq N$ for the first time

N	2	3	4	5	6	7	8	9	10
n	4	11	31	83	227	616	1674	4550	12367

Minimum number of years one needs to wait to see N records (theoretically).

For example, $1/1+1/2+1/3+1/4>2$. From this table, in a random series, on average,

- an eleven-year-old child has seen *three personal* records;
- a mother, thirty-one years old, *four* records;

- a grandmother, eighty-three years young, *five* records.

Isn't this a key to the fact that in our youth, *winters* were colder with more snow? Summers were warmer? Basically, everything was better?

Questions

1. How long would the present record stand (survival time)?
2. What would be the value of the next record?
3. How long should we expect to wait for the rth record to be set?
4. What can we say about the time of the rth record?
5. What can we say about the value of the rth record?
6. Is there an ultimate record?

Bloomsburg Floods

I applied the theory to the floods in Bloomsburg, the city I live in. The probability of a record flood during the next ten years is 0.06. Also, with a 90 percent confidence, the largest possible flood in Bloomsburg will not exceed 33.00.

"Best" Applicant

Suppose that there are <u>one hundred </u>applicants. Probability of one record (first applicant is the best) is $p_{1,100}=0.01$; two records is $p_{2,100}=0.05$. Occurrence of five records has the largest probability, $p_{5,100}=0.21$. So if the manager decides to hire the applicant who creates the *fifth record*, then the probability of hiring the best applicant will be 0.21.

One-Hundred-Meter Dash

During the 2009 world track-and-field competitions, Bolt put on an unbelievable performance, shattering his own record in the one hundred meters, lowering it by 0.11 seconds to an amazing 9.58 seconds.

Question: How long will his record survive?

In most sports, records have occurred more frequently than what the theory predicts.

Men's one-hundred-meter data in 1912–2012 includes twenty records.

To produce twenty records, more than one hundred million independent and identically distributed attempts are needed.

Also, unlike the theoretical expectation, the waiting times have decreased significantly with time especially in the recent past.

To account for these and other contributing factors, we made up for the increase in probability and frequency of the records by inflating the number of attempts.

A Demonstrating Example

Consider the data for the 100m dash. The median number of attempts required to arrive at a new record is $e=(2.718...)$ times the median number of attempts that was required to arrive at the previous record.

This suggests a geometric increase with rate $e=2.718...$ This leads to probability estimates for a new record.

| 0.152461 next year | 0.562681 next 5 years | 0.808753 next 10 years |

Note: one would expect even better results if the geometric increase is replaced by increase in male population of the world.

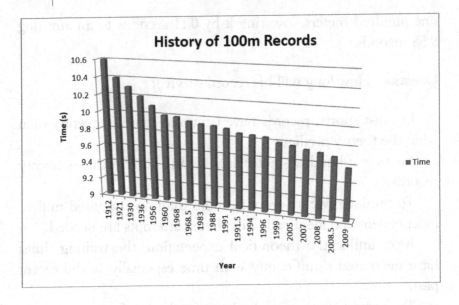

Other Applications of Records

1. Test of randomness. Observing many record highs or lows suggests that the data are not a simple random sample— that is, an alternative hypothesis should be sought to fit the data better.

Of course, it is possible for one hundred random observations to be ordered so that the sequence has as many as one hundred record highs.

But detailed calculation shows that the probability of ten or more record highs in a one-hundred-long random sequence is less than 5 percent.

Formal procedures are available to test the hypothesis of randomness based on the sum or the difference of record-high and record-low frequencies.

2. Car caravans in a one-lane tunnel. When traffic moving in one direction is confined to a single lane, a slow car is likely to be followed closely by a queue of vehicles whose

drivers wish to go faster but who cannot pass. If there is no exit from this lane, then more and more following vehicles will catch up and be added to the slow moving "platoon" or "caravan" until there happens to be following vehicle travelling at a lower speed.

This vehicle will not catch up but will accumulate its own caravan.

Thus, cars whose drivers all desire different speeds, in fact, will travel in caravans at actual speeds determined by record lows in the sequence of desired speeds.

Applying the simple probability model to a random sequence of drivers, the frequency of record lows corresponds to the number of caravans formed by n drivers. And the number of trials between successive record-breaking low values correspond to the lengths of caravans.

Since caravans will be successively slower, separations between caravans will increase as time passes. G. F. Newell used this reasoning to explain why cars near the exit of long tunnel tend to travel faster and in smaller bunches, more widely separated, than cars in the tunnel near the entrance. This model of traffic flow also has been mentioned by other authors.

3. Sequential strategy for destructive testing. We usually teach mathematical methods for finding the minimum of a function. However, we never discuss any strategy for this when dealing with the real-world problems.

Many products fail under stress. For example, a wood beam breaks when sufficient perpendicular force is applied to it, and electronic component ceases to function in an environment of too high temperature; and a battery dies under the stress of time.

But the precise breaking stress or failure point varies even among "identical" items.

Suppose that I can observe item's exact failure point in a laboratory by gradually increasing stress (force, temperature, time, etc.). From such destructive testing of one hundred items, I could find all their failure points, say, $X_1, X_2, ... X_{100}$. But now suppose that I only need to find the weakest item in my sample: I only want the minimum value among failure stress $X_1, X_2, ... X_{100}$. Then I need not stress most of the items to their failure points.

The minimum failure stress among any batch of items can be determined sequentially. Test the first item until it fails and record its failure stress X_1. Stop the next test (short of failure) if the second specimen survives this amount: so the second specimen's failure stress X_2 is determined exactly if $X_2 < X_1$; otherwise, obtain only the "censored" information that $X_2 > X_1$, and hence $X_1 = \max(X_1, X_2)$. Either case, proceed to the third specimen and stop the test if this item survives a stress equal to $\min(X_1, X_2)$: so X_3 is observed only if $X_3 < \min(X_1, X_2)$; but $\min(X_1, X_2, X_3)$ is always determined. In general, the ith item survives its stress test if $X_i > \min(X_1, ... X_i)$ $< \min(X_1, ... X_{i-1})$ or the test concludes with stress-to-failure if $X_i = \min(X_1, ... X_i) < \min(X_1, ... X_{i-1})$. In either case, the value $\min(X_1, ... X_i)$ is known after the i trial.

The items destroyed in this sequential procedure are those with "record low" failure points. The frequency of such record lows fits the same probability model as the low in a sequence of weather records. For a sample of n items, the expected number of items destroyed is $1 + (1/2) + (1/3) + ... + (1/n)$. This harmonic sum grows very slowly compared to sample size n. For example, the sum is only 5.19 when $n = 100$ and is only 7.49 when $n = 1000$.

The sequential strategy to find the minimum value generalizes easily to find the 2,3,..., or j smallest values among $(X_1, X_2, ... X_n)$. To begin, test j items until they fail, at stresses (X_1, X_2, X_j). Thereafter, stop the ith trial if the item survives the j lowest failure stresses among all i - 1 previous specimens. The probability of stress-to-failure for the ith item $(i > j)$ is the probability that it is among the j smallest of i independent observations from the same continuous distribution:

All ranks are equally likely for X_i so the desired probability is j/i. The expected number of items destroyed is the sum of failure probabilities over all trials:

$$1+1+...+1+j/j+1)+j/(j+2)+...+j/n \text{ j terms}$$
$$j[1+1/(j+1)+1/(j+2)+...+1/n]j[1+1/2+1/3+...+1/n]$$

If j is much less than the sample size n, then so is the expected number of failures. For example, to find the weakest four items in a sample of one thousand, I expect to destroy only about twenty-six; and to find the weakest eight items, I expect to destroy less than fifty.

Hiring the Best Applicant/Revisited

A company is looking for a typist. An advertisement was answered by n people, and the company decided to hire one who was the fastest (or any other measure of performance). Each applicant wants to know immediately whether he or she is hired. Applicants who are refused are no longer interested in this job, leave, and will not return. The problem is to find a strategy that ensures the highest probability that the best applicant is hired.

Suggested Solution 1

Let X_i be the typing speed of ith applicant. Assume that applications appear in random order and that all the variables X_i are different. One managerial strategy can be based on the theory of records.

Consider $n=100$, for example. Table indicates that in random order of applicant, the series $X_1, X_2,...X_n, n=100$ has one record with probability $p_{1,100}=0101$, two records with probability $p_{2,100}=0.05$, and so on.

The occurrence of five records has the largest probability, $p_{5,100}=0.21$. If the manager decides to hire the applicant who creates the fifth record, then the probability of hiring the best applicant will be 0.21.

Suggested Solution 2

Another manager's strategy is to examine first s-1 applicants but to hire none of them.

The manager hires the first of the remaining applicants (if that person is present) who has shown better performances than the previous applicants. If such an applicant is not present, the manager must hire the last one. For a given s, we calculate the probability $p(s,n)$ that the best applicant is chosen. Finally, we choose s such that $p(s,n)$ is maximal. With this strategy, the manager examines the first thirty-seven applicants and hires the next one whose test score exceeds that of the preceding applicants. The probability that the best applicant will be chosen is 0.371. If n is large, we can use the Euler formula. Then

$$p(s,n) \approx \frac{s-1}{n} \ln \frac{n-1}{s-2}.$$

For n=10, he should examine three and hire the next one whose test score exceeds that of preceding applicants. For n=5, examine two. For n=15, examine five.

Questions

1. How long would the present record stand?
2. How many records do we expect to observe in n attempts? (n years)
3. How long can we expect to wait for a record to be broken?
4. What would be the value of the new record?
5. How long should we expect to wait for the rth record to be set?
6. What can we say about the value of the rth record?
7. Is there an ultimate record?

A nice feature of the theory of records is that several of these questions can be answered with a nonparametric or distribution free analysis–that is, we do not have to know what the probability distribution of observations is.

Prediction of Records, Short Term

To predict future records, the following approach can be used. For independent and identically distributed observations,

$$1 + 1/2 + \ldots + 1/83 = 5.$$

Long jump data has five records in forty-three years. Assuming that i is the geometric rate of improvement (increase in number of attempts)–that is, every year, the number of attempts has been i time the previous year. We can find its value by solving the equation:

$$1 + i + i^2 + \ldots + i^{42} = 83.$$

Solving for i, we obtain a value of 1.027, which is a rate of improvement 2.7 percent (2.7 percent more attempts per year). For the long jump data, this means that attempts are increased as

1962	1963	1964	2000	2004
1	1.027	$(1.027)^2$	$(1.027)^{38}$	$(1.027)^{42}$

Now, to predict records for the next ten years, we set

$$n_1 = 83, n_2 = (1.027)^{43} + \ldots + (1.027)^{52} = 36.51.$$

For this example, the probability of a new record occurring before 2014 is

$$n_1 / (n_1 + n_2) = 0.31$$

Using the maximum likelihood method, a better estimate for is $n_1 = 73$. This will lead to a smaller value of i.

Ultimate Record

Let $Y_1 > Y_2 > ... > Y_n$ be the order of statistics of distances in the men's long jump.

Then it can be shown that a level 1-p confidence interval for R_u (ultimate record) is given by

$$\{Y_1 - (Y_2 - Y_1) / [(1-p)^{-\alpha} - 1], Y_1\}$$
$$\alpha = \log_{k(n)} / \log [(Y_{k(n)} - Y_3) / (Y_3 - Y_2)]$$

$k(n)$ is any sequence such that $k(n) \to \infty$ and $k(n)/n \to 0$. One natural choice is $k(n) = \sqrt{n}$.

It can be shown that a better choice is

$$k(n) = (eT_r)^{1/2} + (t_r)^{1/2}$$

T_r: Time between the last and previous ultimate record
t_r: Time the last record has held to date

To demonstrate, suppose that we are in the year 2004. For long jump

$Y_1 = 8.95$ (1991) $t_r = 2004 - 1991 = 10$
$Y_2 = 8.90$ (1968) $T_r = 1991 - 1968 = 23$
$Y_3 = 8.86$ (year 1987) $Y_{11} = 8.63$ (year 1997)
$k(n) = (eT_r)^{1/2} + (t_r)^{1/2} = 11$
$\alpha = \ln(k(n)) / \ln [(Y_{k(n)} - Y_3) / (Y_3 - Y_2)] = \ln 11 / \ln [(8.63 - 8.86) / (8.86 - 8.90)] = 1.37085$
$Y_1 - (Y_2 - Y_1) / [(1-p)^{-\alpha} - 1] = 8.95 - (8.90 - 8.95) / [(0.95)^{-1.37085} - 1] = 9.64$

With 95 percent confidence, the ultimate record for men's long jump will lie within the interval [8.95, 9.64].

Alternative Short-Term Prediction

In year 2005, I managed to develop a method utilizing the following three results of the theory of records for independent and identically distributed sequence of observations:

a) If there is an initial sequence of n_1 observations and a batch of n_2 future observations, then the probability for this additional batch to contain a new record is $n_2/(n_1+n_2)$.

(B) As sample size $n \to \infty$, the frequency of the records among observations indexed by $an < i < bn$ tends to a Poisson count with mean $ln(b/a)$.

c) If $F(y)=1-e^y, y>0$ and Y_N denote the record values and if $D_1 = Y_{N_1}, D_r = Y_{N_R} - Y_{N_{r-1}}, r \geq 2$, then the improvements $D_1 D_2$ ... are independent and identically distributed (iid) random variables with common distribution function $F(y)$.

Clearly, the results of theory of records for independent and identically distributed sequences are not suitable for sports since sport records are usually more frequent than what the theory predicts. To account for this, some adjustment is therefore necessary. Noubary has treated the problem as if either participation has increased with time or more competitions have taken place so that the chance of setting a new record was increased.

For Boston marathon, the participation has steadily increased during the years. Usually, 1970 is selected as the starting year because during this year, a qualifying time was introduced for participation. One simple approach would be to model the increase and use that together with result (a) above for prediction. For example, using regression, the model given below was found for the number of participants as a function of the year ($R^2 = 0.938$).

Number of participants in year
$$t=-1294+1088t-57.5t^2+1.25t^3$$

In this model, the data for the year 1996 was replaced by the average of the two neighboring values since this was the one-hundredth running of the Boston marathon and more than thirty-eight thousand runners were allowed to participate. Now, as this approach seems reasonable as it could account for other factors such as advanced training programs, better equipment, diet, coaching, and even use of steroids, all of which increase the chance of setting new records.

To clarify, suppose that it is now year 1924 and the best record for men's one-hundred-meter run is b and the probability of breaking this record is p. Suppose further that in year 2000, the population of word has tripled. If we divide this population to three groups, then each group could break the record with probability p.

Thus, the probability that at least one group breaks the record is $1-(1-p)^3$. For example, this probability is 0.143 for p-0.05.

Consider the following exponential model for the growth of the world's male population since 1900.

$$\text{Population in year } t = 1.6 \exp{[.0088(t-1900)]}$$

Note that this means a geometric increase of $\exp(0.0088)-1=.00884$ per year since year 1900.

We can use this assuming that the number of participants or attempts is proportional to the population size at time t. However, this approach does not use information from the sports itself and the way records were set. In other words, it is the same for all sports regardless. We can also apply result *(b)* assuming a geometric increase.

According to this result, the frequency of the records among observations 84 and 137 (sum of 84 and 53) has approximately a Poisson distribution with mean

$$\lambda = \ln(137/84) = 0.489.$$

Using this, the probabilities of zero and one record in the period 2000–2009 are, respectively, 0.613 and 0.300. Here, 1−0.613=0.387 is an estimate for probability of at least one record in the next ten years. Finally, let us demonstrate application of the result *(c)*.

The probability for occurrence of a record larger than m_0, say, in the next n_2 years, can be calculated using the following formula obtained by combining results *(a)* and *(b)*.

$$P(m > m_0) = \frac{n_2}{n_1 + n_2} \exp(-(m_0 - 8.95)/0.195)$$

Note that, here, 8.95 is the value of the last (fifth) record.

As an example, for $m_0 = 9$, $P(m>m_0)$ is, respectively, equal to 0.0329 and 0.3024 for the future one and ten years, assuming a geometric increase.

To account for these and other contributing factors such as diet, shoes, track type, etc., we treat the problem as an independent and identically distributed one, but make up for the increase in probability and frequency of records by inflating the number of attempts. We start by stating results of the theory of records for independent and identically distributed observations that we plan to use:

(a) If there is an initial sequence of n_1 observations and a batch of n_2 future observations, then the probability that the additional batch contains a new record is $n_2/(n_1+n_2)$.
(b) For large $n, P_{r,n}$, the probability that a series of length n contains exactly r records is given by

$$P_{r,n} \quad \frac{1}{(r-1)!n}(\ln(n) + y)^{r-1},$$

where $y=0.5772$ is Euler's constant (see appendix).

(C) As sample size $n \to \infty$, the frequency of the records among observations indexed by $an < i < bn$ tends to a Poisson count with mean $ln(b/a)$.

(d) The median of W_r, the waiting time between the $(r-1)^{th}$ and r^{th} records, are

Record Number r	2	3	4	5	6	7	8
Median (W_r)	4	10	26	69	183	490	1316
Med (W_r)/Med (W_{r-1})		2.50	2.60	2.65	2.65	2.68	2.69

Medians of Waiting Times between Successive Records and Their Ratios

Moreover,

$$\frac{\text{Median}(W_{r+1})}{\text{Median}(W_r)} \approx e = 2.718\ldots$$

Application to One-Hundred-Meter Data

Using the maximum likelihood method and maximizing $P_{r,n}$ with respect to n, we find $n = 100,212,150$. $100,212,150$ is an estimate for the number of iid attempts that is required to produce twenty records. Next, we need to distribute these attempts over the period of one hundred years using an increasing function or pattern. The median number of attempts required to arrive at a new record is $e = (2.718\ldots)$times the median number of attempts that was required to arrive at the previous record. This suggests a *geometric increase* with rate $e = 2.718\ldots$

If we assume that *one unit of attempt* was needed to arrive at the second record, the total number of attempts to arrive at record number twenty may be calculated as

$$1 + e + e^2 + \ldots + e^{18} = 103,872,541.$$

This is a slight overestimation, as for early records, the ratios are less than e.

Suppose i is the annual *geometric rate* of increase in number of attempts.

The number of attempts in year k is i times the number of attempts in year $(k-1)$.

Then i can be found by solving the equation

$$1 + i + i^2 + \ldots + i^{100} = 100,212,150$$

i=1.17988–that is, 17.9888 percent more attempts per year. This leads to probability estimates for a new record 0.152461 next year, 0.562681 next five years, and 0.808753 next ten years.

Note: one would expect even better results if the geometric increase is replaced by increase in male population of the world could.

Models such as Logistic or Gompertz or, more generally, a model of the form

$$y_{n+1} - y_n = H(y_n) = i * f(y_n)(1 - g(y_n)).$$

y_n: number of participants or number of attempts at year n

Noubary (2005) considered the following simpler model

$$y_{n+1} = y_n exp[r * (1 - y_n / h)]$$

For example, the number of attempts in the future one and ten years are 412 and 4876 using y_0=100, r^*=0.04 and h=50. The corresponding numbers using the logistic equation are 402 and 4742. These numbers result in smaller probability estimates compared to the geometric increase of 4 percent.

More on Ultimate Records

Although ultimate records in any area of application are of great importance, for simplicity here we concentrate on sport records. Sports provide an inexhaustible source of fascinating and challenging problems in many disciplines including mathematics. According to authoritative opinion, modern sports have been getting more and more intellectualized, and more advanced scientific analysis is presented. At the same time, extra attention is paid to the fact that it is possible to study many situations in sports from a mathematical perspective and that it is desirable to obtain more valid quantitative and qualitative estimates of the things happening in sports. For example, mathematical methods are applied to estimate an athlete's chances of success, identify the best training conditions for him/her, and measure his/her effectiveness. Information theory makes it possible to estimate the amount of eyestrain in mountain skiing, table tennis, and so on. Mathematical physics is used to identify the best shape of rowboats and oars. In general, applied probability and statistics has been instrumental in analysis of sport data. For example, being profitable, baseball has long since been an object of attention for sport and business interests. A vast volume of statistics has been accumulated, enabling experts to draw conclusions about the quality of a team's performance (the average number of successful pitches, depending on the pitcher's and catcher's proficiency, the law of distribution of hits, and so on). A simulation model of baseball was constructed with the help of the probabilistic Monte Carlo method. Soon after, mathematical methods were applied to football. One paper contains the analysis of 8,373 games in fifty-six rounds, including the U.S. National Football League table. It supplies important recommendations on offensive strategy.

In sum, athletic competitions provide the researcher with a wealth of material that is registered by coaches, carefully preserved and continuously piled up. There is plenty of opportunity out there to experiment, to test mathematical models and optimal strategies in situations occurring in sports. Only a tiny part–quite possibly not the most intriguing one–of the problems arising in sports has been described in the pages of books and journals. Think how many yet unsolved problems arise in different sports.

We note that apart from intrinsic interest, there are several medical and physiological reasons why we would like to know how fast a human being could, for example, run a short distance such as one-hundred-meter dash or a medium distance such as four hundred meters. While there is a general agreement among the physiologists and physical educators about existence of an upper limit for such a speed (or a lower limit for the time needed to run this distance), the limit is not known at the present time. Because of a great interest in this question, apart from physiological research, there have been some attempts to estimate (predict) the limits via statistical modeling.

Analyzing and Communicating Flood Risk

Property and business owners as well as their elected or appointed decision makers in small, rural, or isolated communities do not have access to the type of data, maps, or interpretive methodologies frequently cited as "best practices" by experts. They do, however, generally record and aggregate data regarding the date and height of floodwater on a point-by-point basis. We propose to leverage this data, where available, by using one or more well-known statistical tools that can be sourced by project managers and others to improve understanding, collaboration, and decision-making. Application to the data derived from the Susquehanna River in the vicinity of Bloomsburg, Pennsylvania, suggests that these tools can succinctly address point-specific issues in a simple, familiar framework to streamline the planning process. Alternative analyses such as these may be valuable to engineers and others with academic, professional, and humanitarian interests in developing countries or regions.

Introduction

Periodic floods have a major influence on the scope and quality of lowlands (650 feet elevation or lower), which harbor most of the world's most populous cities. To make matters worse, cities and other urban areas are even more prone to flooding than rural environments because of the relatively greater area covered by pavement and other impermeable materials that limit percolation, hasten runoff, and increase risk. Despite this, people are increasingly attracted to urban areas because of the availability of employment, education, health services, and a variety of other cultural and economic factors. In fact, it is estimated that global flood damage could exceed $1 trillion annually by 2050.

On a worldwide basis, government has largely failed to solve this problem. Rapid development continues in flood-prone areas throughout the world, and people as well as their investments remain at risk. In many cases, the capacities of existing flood control measures such as dams, seawalls, and other barriers are already woefully inadequate given the loss of natural percolation areas that rapid urbanization brings. Even if the funding could be found to build new flood control infrastructure, the costs would be staggering.

It seems to us that the prudent and responsible alternative would be to limit development in flood-prone areas to agricultural, park land, or other uses where periodic flooding would not cause significant economic or social impacts.

However, decision makers are often at a loss to evaluate actual flood risk. Technical studies are often prohibitively expensive and take a very long time to complete, but there are alternatives. Whereas most small, isolated communities do not have access to sophisticated studies, current maps, and other best practice tools (see U.S. Army Corps of Engineers How to Communicate Risk [http://www.corpsriskanalysisgateway.us/riskcom-toolbox-communicate.cfm]), they do generally collect and aggregate the date and height of floodwater.

In this paper, we propose a few alternative methods to evaluate flood risk in smaller or more-marginalized communities that do not have updated maps or more-sophisticated means to analyze or interpret their records. These alternatives are based on statistical concepts that leverage simple data like water height and date of flood events. The methods proposed involve the analysis of extremes, exceedances, excesses, record heights, and ultimate heights of floodwater.

To demonstrate, methods are applied to data recorded since 1850 from the Susquehanna River at the town of Bloomsburg in central Pennsylvania. Bloomsburg offers a particularly valuable opportunity to test these tools in that the data have been accurately and meticulously recorded for over 150 years. Further, floods are a significant problem in the area, and they compromise critical infrastructure such as drinking water, sewer, transportation, and emergency services.

The river in Bloomsburg floods when the water level exceeds 19 feet. Since 1850, there have been thirty-eight floods, with 32.75 feet establishing the highest recorded water level. We hope by using the values derived from these data, we offer a useful and "user friendly" platform to consider flood risk to decision makers and their constituents in small towns like Bloomsburg and elsewhere in the world especially where flooding can have catastrophic consequences.

Empirical Rule

For the Bloomsburg data, the mean, median, and standard deviation are, respectively, 24.3 feet, 23.4 feet, and 3.4. Also, the percentages of floods with sizes to within one, two, and three standard deviations from the mean are, respectively, 71, 92, and 100 percent. These percentages agree with the empirical rule indicating the normality this data set for. Thus, the probabilities of a future flood greater than 30.9 feet or 32.1 feet are, respectively, 5 percent and 1 percent. Also, the probability of a record flood (greater than 32.75 feet) is 0.006. This indicates that there is a small chance that a randomly selected future flood cresting level will exceed the present record.

Exceedances

The theory of exceedances deals with the number of times a specified threshold is exceeded. Assuming independent and identically distributed events (floods), we may wish to determine the probability of r exceedances in the next n occurrences. To apply this to Bloomsburg data, we note that since 1850, there have been thirty-eight floods exceeding 19.8 feet, with the largest 32.7 feet. So the mean expected value, variance, and standard deviation of the number of exceedances of 32.7 feet during the next ten and one hundred years would, respectively, be

$10/173 = 0.058$, (10) (172) (183)/ (173)2(174) = 0.06, and 0.246
$100/173 = 0.58$, (100) (172) (273)/ (173)2(174) = 0.902, and 0.95

These estimates have relatively large standard deviations because of a small sample size.

Records

This theory deals with values that are strictly greater than or less than all previous values. Usually, the first value is counted as a "record." A value is a record (upper record or record high) if it exceeds or is superior to all previous values.

To predict the future records, we have developed a simple method utilizing the following results of the theory of records for an independent and identically distributed sequence of observations (McDonnell 2013).

(a) If there is an initial sequence of n_1 observations and a batch of n_2 future observations, then the probability that this additional batch contains a new record is $n_2/(n_1+n_2)$.

(b) For large n, $P_{r,n}$, the probability that a series of length n contains exactly r records is

$P_{r,n}$ is approximately $\dfrac{1}{(r-1)!\,n}(\ln(n)+y)^{r-1}$.

For the Bloomsburg data, we note that there have been thirty-eight floods exceeding 19.8 feet. This gives a rate of 38/166 per year. Thus, the probability estimate of a record flood (as in [a] above) during the next ten years is

$$10/176 = 0.057 \sim 0.06.$$

Also, using (b), the probability of three records in thirty-eight observations is 0.003. With the above rate, we expect two floods in the next ten years, and the probability of three records in forty observations is 0.002930091. Hence, P (no record in next 10 years) = P (three records in forty observations) / (three records in the thirty-eight observations) = P (three records in forty observations) / P (three records in thirty-eight observations) = 0.94. This gives the probability of a record in the next ten years as 0.06, or 6 percent.

Excesses

In this approach, the probabilities of future large floods are calculated by developing models for the upper tail of the distribution for height of floodwaters. Because values above an appropriate threshold carry more information about the future large floods, this approach is reasonable. Here, one usually assumes that the tail of the distribution for flood sizes belongs to a given parametric family and then proceeds to do inference using excesses–that is, the floods greater than some predetermined value. It has been shown that the most appropriate model for tail is the so-called Pareto distribution, which includes models for short, medium, and long tails. We applied this approach to the Bloomsburg data and found that the best fit is a short tail based on the largest four floods as follows:

$$\overline{G}_l(x) = 1 - (1 + 0.24327x)^{-0.15764} \quad 0 < x < 4.1106.$$

This led to the following upper bound for floods in Bloomsburg:

$$V_{max} = 32.7 \text{ ft.}$$

This is virtually the same as the largest flood in Bloomsburg that occurred in 2011.

Ultimate Flood

In terms of predictive value, we can avoid large standard errors and provide a confidence interval for the upper bound based on the most recent large floods or record floods. Let Y and $Y_1, Y_2, ..., Y_n$ represent flood size for a given region, where

$$Y_1 \leq Y_2 \leq ... \leq Y_n.$$

Assuming that the distribution function $F(y)$ has a lower endpoint and certain conditions are satisfied, a level $(1-p)$ confidence interval for the maximum of Y is (De Haan 1981)

$$\{Y_1 + (Y_2 - Y_1)/[(1-p)^{-k} - 1], Y_1\}.$$

De Haan (1981) has also shown that

$$\frac{\ln m(n)}{\ln[(y_{m(n)} - y_3)/(y_3 - y_2)]}$$

is a good estimate for k.

To apply this result, we need to choose an integer $m(n)$ depending on n such that $m(n) \to \infty$ and $m(n)/n \to 0$ as $n \to \infty$. It is shown that the following choice works well even for the worst case:

$$m(n) = \sqrt{eT_r} + \sqrt{t_r} = \sqrt{2.718282T_r} + \sqrt{t_r}$$

For the Bloomsburg data, $T_r = 107$ and $t_r = 1$ and $m(n)=18$. Since $y_1=32.75$ feet, $y_2=32.70$ feet, $y_3=31.2$ feet, $y_{18}=23.5$ feet. We have $k = 1.767$ leading to a 90 percent one-sided prediction interval for the upper bound as (32.75 feet, 33.0 feet).

Summary

The application of alternative statistical methodology demonstrated in this paper for the Susquehanna/Bloomsburg area is a useful complement to best practice methods to evaluate flood risk in that

1. they can be used to communicate point-specific risk and answer the question most commonly asked–"How likely is it that this particular area will be inundated in any given year?";
2. risk can be communicated using a familiar, simple scale such as "On a scale of one to ten, this area scores eight or has an 80 percent chance of flooding in any given year";
3. threat can be evaluated quickly with relevant data and without immediate reference to maps, which may not be available; and
4. such analyses can serve as an important touchstone between decision makers, developers, and others early and effectively in the flood control process.

Instances of flooding (water level >19.0 feet) at a point of reference in the vicinity of Bloomsburg, Pennsylvania.

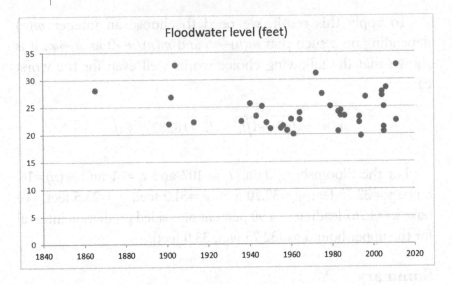

Concluding Remarks

Although the number of floods has increased significantly over time, the average strength and height of these floods remain constant and independent of their time of occurrence. This increase in frequency of floods over time should be of great concern as it leads to a hazardous condition. If flood frequency continues to increase at such a rate, they can become a yearly occurrence that could damage towns and destroy communities. Measures need to be taken to save areas near rivers and lakes.

References

Arensburg, A., and S. Hutt. 2007. "Challenges and Issues for Water Management in Northwest China." *IDEAS*.

Brenner, R., C. Keung, B. Rosenblum, R. Soltz, and S. Wolfe. 2013. "China's Loess Plateau–A Region of Heterogeneous Environmental Communities." Proc., GASI, Rome, Italy.

Davis, R. and S. Resnick. 1984. "Tail Estimates Motivated by Extreme-Value Theory." *Annals of Statistics* 12: 1467-1487.

De Haan, L. 1981. "Estimation of the Minimum of a Function Using Order Statistics." *Journal of the American Statistical Association* 76: 467-469.

Department of the Army, Office of the Chief Engineer. EM 1110-2-1411, March 1965.

Federal Emergency Management Agency [FEMA], Region 10. The 100-year Flood Myth. Pages 1-6, undated training document.

Hill, B. M. 1975. "A Simple General Approach to Inference about the Tail of a Distribution." *Annals of Statistics* 3: 1163-1174.

McDonnell, T. 2013. "Global Flood Damage." *Mother Jones.*

Pickands, J. 1975. "Statistical Inference Using Extreme Order Statistics." *Annals of Statistics* 3: 131-199.

Smith, R. L. 1988. "Forecasting Records by Maximum Likelihood." *J. of the Amer. Stat. Assoc.* 83: 331-338.

Air Quality Data

Extreme pollutant concentration is of great concern because of its bad effect. This paper presents a technique for the statistical modeling of sample extremes, which come as excesses over a threshold, and demonstrates its application for the prediction of extreme pollutant concentrations. The problem of statistical inference for such excesses may also arise in any area of science where the analysis of extremes of observation sequences is important, for example, in hydrology, meteorology, oceanography, and seismology.

The distribution function of the mean concentration of air pollutants has been studied and described by many authors, and it appears that a lognormal distribution would give a reasonable fit, at least in the center of distribution. However, it has also been demonstrated that the often-used assumption of log-normality is unreasonable and that divergence, especially in the higher percentiles, increases with averaging time.

Considering the large values, a number of approaches to modeling upper extremes of such sequences may be possible, depending on the structure and complexity of the data. If the sequences are fairly long, the classical method of treating annual maxima of consecutive periods of equal length, for example, years, months, or days of the series as independently and identically distributed in one of the extreme value distributions, is often used following the work of Gumbel (1958).

In the case of air pollution data, Gumbel's type I asymptotic distribution of maxima is generally used by investigators (see Roberts 1979a, b, and Surman et al. 1986). One reason for this is that the lognormal distribution belongs to the domain of attraction of the type I distribution. However, one should note that a good fit to log-normality in the center of the distribution does not ensure that the type I distribution will fit the maxima well.

Table I. Sulfur dioxide, 1 h average concentrations (pphm); monthly and annual maxima

Year	Jan.	Feb.	Mar.	Apr.	May	Jun.	Jul.	Aug.	Sep.	Oct.	Nov.	Dec.	Max.
1956	47	31	44	12	13	3	14	21	33	26	40	32	47
1957	22	19	20	32	20	23	18	16	13	14	41	25	42
1958	15	13	20	12	24	13	37	20	32	27	27	68	68
1959	20	32	20	15	3	6	8	15	17	15	20	20	32
1960	22	18	23	20	8	13	14	9	13	16	27	20	27
1961	25	20	20	16	10	10	8	10	12	16	14	43	43
1962	20	13	15	18	10	12	10	10	11	11	14	7	20
1963	12	18	27	21	2	7	4	4	15	10	18	18	27
1964	16	10	3	3	19	9	16	25	4	14	18	21	25
1965	16	18	9	14	8	10	18	18	14	12	17	14	18
1966	27	33	25	10	17	30	13	18	22	15	25	23	33
1967	30	40	32	10	8	7	8	26	10	40	18	17	40
1968	51	30	18	22	10	19	22	25	26	29	50	40	51
1969	37	13	55	14	9	10	13	17	33	13	15	44	55
1970	23	19	10	11	15	12	25	40	25	20	15	8	40
1971	22	36	20	28	10	15	20	55	38	41	26	25	55

1972	30	32	18	27	37	13	23	19	21	31	25	13	37
1973	10	8	8	12	11	16	25	16	11	28	10	23	28
1974	8	9	9	13	8	14	9	9	25	11	19	15	25
Average:	23.8	21.6	20.8	16.3	12.7	13.7	16.0	19.6	19.7	20.4	22.9	25.0	

Table II. Order statistics of annual maxima and the twenty largest monthly maxima

Ann.	68	55	55	51	47	43	41	40	40	37	33	32	28	27	27	25	25	20	18	–
Mon.	68	55	55	51	50	47	44	44	43	41	41	40	40	40	40	40	38	37	37	37

Now, although Gumbel's method is common and often successful in environmental applications, it has some drawbacks:

(i) It is not clear which of three possible types should be used.
(ii) More seriously, its use of data is rather uneconomical, and inferences based on short sequences are often unreliable. This last difficulty is common to all methods of analysis for sample extremes; the point here is that if, say, k years of data are available, then inference based on the upper k order statistics (largest values) ought to be at least as good as inference based on the annual maxima. This is because order statistics are at least as great as the annual maxima and so might be expected to be more informative about the upper tail of the distribution.

To clarify (ii), consider, for example, the data from nineteen years' observations of 1 h mean concentrations from Long Beach, California (taken from Roberts 1979b) presented in table 1. Suppose, as usual, we are considering the annual maxima so that we have nineteen observations. The order statistics corresponding to this is given in the first row of the table 2. To make a comparison, the nineteen largest observations of monthly maxima are also presented in row 2. As can be seen, several large values are ruled out in yearly maxima. Thus, if inference concerning the upper tail is of main interest, consideration of annual maxima only is an inefficient utilization of the data. In other words, a great deal of information available is ignored.

Method

We now describe a new method to make inference about the upper tail of a distribution that, as we shall see, makes use of all the large observations in the sample. The method, known as the "threshold method," had been introduced and used by hydrologists in its earliest form and later developed from a more rigorous mathematical standpoint by statisticians.

The basic idea is that the observations are plotted in order of time, and the excesses above an arbitrarily chosen threshold (denoted by u) are noted. The excess time $T_1, T_2, \ldots$ are the times at which the excesses occur, and the excesses $Y_1, Y_2, \ldots$ are the differences between the observations over the threshold and the threshold itself. Analysis proceeds by fitting appropriate statistical models to the T's and the Y's.

The excess times may be thought of as a point process on the real line for which appropriate models are either homogeneous or nonhomogeneous Poisson processes. Statistical methods for fitting models of this type are well established (see Cox and Lewis 1966). Let us consider the excesses $Y_1, Y_2, \ldots$ which are our main concern. For this, it is shown that the natural parametric family of distribution to consider is the so-called generalized Pareto distribution (henceforth GPD)—namely,

$$H(y;\sigma,k) = 1 - 1\left(1 - k\frac{y}{\sigma}\right)^{1/k}.$$

Here, $\sigma > 0$, $-\infty < k < \infty$, the range of y is $0 < y < \infty$ $(k \leq 0)$, $0 < y < \sigma/k$ $(k > 0)$. In fact, this is motivated by the following considerations:

(1) The GPD arises as a class of limit distributions for the excess over a threshold, as the threshold increases toward the right-hand end of the distribution (i.e., tail). This property, discovered by Pickands (1975), is a direct analogue of the classical limit theory, which was used to justify the extreme value distributions.

(2) If Y has the distribution function $H(y, \sigma)$ and $y' > 0$, $\sigma - ky' > 0$, then the conditional distribution function of $Y - y'$ given $Y > y'$ is $H(y; \sigma - ky', k)$. This is a "threshold stability" property; if the threshold is increased by an arbitrary amount y', then the GPD form of the distribution is unchanged. Again, this is a direct analogue of a classical property of the extreme value distributions.

(3) Here is a more direct connection between the GPD and the extreme value distributions. If N is a Poisson random variable with mean λ and $Y_1, Y_2, \ldots, Y_N$ are independent excesses with the distribution function $H(y; \sigma, k)$, then

$$P(\max(Y_1, Y_2, \ldots, Y_N) \leq x) = \exp\{-\lambda(1 - kx/\sigma)^{1/k}\},$$

which is the generalized extreme value distribution. Thus, if N denotes the number of excesses in, say, a year and Y_1, $Y_2, \ldots, Y_N$ denote the excesses, then the annual maximum has one of the classical extreme value distributions.

(4) The limit $k \to 0$ of the GPD is exponential distribution. Thus, the earlier forms of the method used by hydrologists may be considered as special cases of this.

In more detail, the proposed method is to treat the excesses as independent random variables and to fit the GPD to them. Dependence on seasonal or other external factors may be incorporated into the parameters σ and for k. Parameter estimation may be carried out by maximum likelihood, graphical, or nonparametric methods. Note that the choice of threshold is, to a large extent, a matter of judgment, although some guidelines can be stated. If it is too low, then the GPD will be a poor fit to the data and the cluster maxima will be too close together to be treated as independent. If it is too high, then there will simply not be enough data above the threshold to obtain reliable estimates. DuMouchel (1983) has proposed the 90th percentile of the sample as a reasonable choice. There are also some data-analytic techniques (see Hill 1975).

Application to Air Pollution Data

To demonstrate the application of the threshold method, let us once more consider the data presented in table 1. For this data, Roberts (1979a, b) has fitted type I extreme value distributions to both monthly and yearly maxima. The results are the following distributions:

$$F_M(y) = \exp[-\exp(-0.115(y - 14.5))],$$
$$F_Y(y) = \exp(-\exp(-0.081(y - 31.5)))].$$

As mentioned before, by using only the maximum of any one year, useful data are discarded. On the other hand, when monthly maxima are fitted, we would have a large number of data (twelve times the yearly maxima), and any adjustment for small or even moderate values will have an effect on the tail and will, therefore, significantly change the respective probabilities. In fact, if large values are of concern, we should base our inference on the large values of our sample.

Turning to the question of threshold, it is clear that a level with a direct physical interpretation should be chosen where possible, provided, of course, that a useful model results. Here, we have taken the twenty largest observations corresponding to threshold 37, which is roughly 10 percent of the data (as suggested by DuMouchel 1983). It is also close to a number of annual maxima (see table 2).

Before proceeding further, let us briefly describe the important estimating procedures available for the GPD. We start with the graphical method because of Davison (1984). Davison observed that if $Y \sim \text{GPD}(\sigma, k)$ and $u > 0$, then

$$E[Y - u \mid Y > u] = \frac{\sigma - ku}{1 + k},$$

provided that $k > -1$. Hence, a plot of (mean excess over u) vs. u should be roughly linear with slope $-k/(1 + k)$ and intercept $\sigma/(1 + k)$. Thus, the plot provides quick estimates of the GPD parameters, and its linearity is a check on the appropriateness of the GPD.

Next, we consider the method based on order statistics presented by Pickands (1975). Suppose the N excesses M_j are arranged in descending order so that $M_1 > M_2 > ... > M_N$. Pickands had taken M_{4h}, where h is the integral part of $N/4$, as the threshold. He then treated $M_1 - M_{4h}, M_2 - M_{4h}, ..., M_{4h-1} - M_{4h}$ as though they were the descending ordered statistics of a sample of size $4h - 1$ from a GPD. The parameters were then estimated as

$$\tilde{k} = (\log 2)^{-1} \log[(M_{2h} - M_{4h}) / (M_h - M_{2h})]$$

and

$$\tilde{\sigma} = -\tilde{k}(M_{2h} - M_{4h}) / (2^{\tilde{k}} - 1).$$

Finally, to obtain the maximum likelihood estimates, we first reparametrize the problem in terms of ω and k, where $\omega = k/\sigma$. Then the maximum likelihood estimators for k, conditionally on ω is given in closed form by

$$\hat{k}(\omega) = -\frac{1}{N} \sum_{j=1}^{N} \log(1 - \omega M j),$$

where $M_1, M_2, ..., M_N$ are the excesses. The likelihood equation for the w then becomes $\Phi(\omega) = 0$, where

$$\Phi(\omega) = \frac{1}{\omega} - \frac{1}{N}\left[\frac{1}{k(\omega)} - 1\right]\sum_{j=1}^{N}\frac{M_j}{1-\omega M_j}.$$

The function $\Phi(\omega)$ is defined for $-\infty < \omega < 1/M^*_N$, where $M^*_N = max(M_1, M_2, ..., M_N)$.

It is also continuous at 0 with

$$\Phi(0) = -\frac{1}{2N\overline{M}}\left(\sum_{j=1}^{N}M_j^2 - 2N\overline{M}^2\right),$$

where

$$\overline{M} = \frac{1}{N}\sum_{j=1}^{N}M_j.$$

Like the generalized extreme-value distribution, the maximum likelihood works only for $k < 1/2$ (see Hall 1982 and DuMouchel 1983). It is shown that, in this case, the asymptotic normal distribution of the estimates based on a sample of size N has a mean (k, σ) and covariance matrix

$$\frac{1}{N}\begin{bmatrix} (1-k)^2 & \sigma(1-k) \\ \sigma(1-k) & 2\sigma^2(1-k) \end{bmatrix}.$$

Note that when $k < 1/2$, one can also set up confidence intervals for the estimated parameters.

Now, to fit the GPD to the twenty largest observations, we first applied the graphical method to see if the maximum likelihood method is applicable here. This was the case, and the result provided us with an initial value for solving the likelihood equations. The final result was the following model for the tail.

$$H(y) = 1 - (1 + 0.3125(y - 37))^{-0.8546}, y \geq 37.$$

We now calculate some probabilities. First, the probability that the next monthly maximum is higher than $\alpha > 37$ is

$$P(y \geq \alpha) = (20/229)(1 - H(\alpha)),$$

where 20/229 is the expected value of the area under the parent probability density function for a 1 h average concentration. (In general, the expected value of the area under a probability density function between two successive order statistics corresponding to a random sample of size N is equal to $1/(N + 1)$.) Also, the probability of having a value higher than α in anyone year is

$$P(y \geq \alpha) = 1 - (1 - (20/229)(1 - H(\alpha)))^{12}$$

Next, to make comparison, we have considered $\alpha = 50$. This is because the United States Federal short-term standard for sulfur dioxide (SO_2) requires that the 3 h mean concentration of SO_2 should not exceed 50 pphm (parts per hundred million) more than once a year. Since the 3 h means have less accentuated extremes than the 1 h means, our procedure will certainly not underestimate the frequency of exceedance. For tail model $H(y)$ and extreme value distribution for monthly maxima, $F_M(Y)$, the probabilities of exceedance are, respectively,

$$(20/229)(1 - H(50)) = 0.2183 \text{ and } 1 - F_M(50) = 0.01672,$$

while the observed frequency is $5/228 = 0.02193$. Also (as another example), these are for $\alpha = 42$, respectively,

$$(20/229)(1 - H(42)) = 0.03907 \text{ and } 1 - F_M(42) = 0.04144,$$

while the observed frequency is $9/228 = 0.03947$. Further, for annual exceedance we have

$$1 - (1 - (20/229)(1 - H(50)))^{12} = 0.23273,$$
$$1 - F_Y(50) = 0.20359,$$

while the observed frequency is $5/19 = 0.26316$. As can be seen, in all cases, the values provided by the new model are much closer to the observed frequencies.

Conclusion

By using only the maximum concentrations in any year, some data useful for the prediction of future large values are discarded. On the other hand, the inclusion of a large number of observations, such as monthly maxima, in the analysis, will lead to an inaccurate estimation of the tail and the respective probabilities. The threshold method described here overcomes these difficulties. It has a firm mathematical foundation and is intimately related to the classical extreme value theory. No matter what the underlying distribution is for concentrations, the distribution of the high values can take one of three possible forms. Here, unlike the classical extreme value methods, no intuitive decision is necessary in advance in favor of any one of these three forms.

References

Cox, D. R., and P. A. W. Lewis. 1966. *The Statistical Analysis of Series of Events*. London: Chapman and Hall.

Davidson, A. 1984. "Modelling Excesses over High Thresholds, with an Application." In J. Tiago de Oliveria (ed.), *Statistical Extremes and Application*, D. Reidel, Dordrecht, 461-482.

DuMouchel, W. 1983. "Estimating the Stable Index a in Order to Measure Tail Thickness." *Ann. Statist* 1: 1019-1036.

Gumbel, E. J. 1958. *Statistics of Extremes.* New York: Columbia University Press.

Hall, P. 1982. "On Estimating the Endpoint of a Distribution." *Ann. Statist.* 10: 556-568.

Hill, B. M. 1975. "A Simple General Approach to Inference about the Tail of a Distribution." *Ann. Statist* 3: 1163-1174.

Pickands, J. 1975. "Statistical Inference Using Extreme Order Statistics." *Ann. Statist.* 3: 119-31.

Roberts, F. M. 1979a. "Review of Statistics of Extreme Values with Applications to Air Quality data, Part I." Review. *J. Air. Pollut. Control. Assoc.* 29: 632-637.

Roberts, F. M. 1979b. "Review of Statistics of Extreme Values with Applications to Air Quality data, Part II." Applications. *J. Air Pollut. Control. Assoc.* 29: 733-740.

Surman, P. G., J. Bodero, and R. W. Simpson. 1986. "Application of Extreme Value Theory to Air Quality Data." Pacific Statistical Congress. North-Holland, Amsterdam, 299-302.

Printed in the United States
By Bookmasters